Simulations of Groundwater Flow, Transport, and Age in Albuquerque, New Mexico, for a Study of Transport of Anthropogenic and Natural Contaminants (TANC) to Public-Supply Wells

By Charles E. Heywood

National Water-Quality Assessment Program

Scientific Investigations Report 2012–5242

U.S. Department of the Interior
U.S. Geological Survey

U.S. Department of the Interior
SALLY JEWELL, Secretary

U.S. Geological Survey
Suzette M. Kimball, Acting Director

U.S. Geological Survey, Reston, Virginia: 2013

For more information on the USGS—the Federal source for science about the Earth, its natural and living resources, natural hazards, and the environment, visit http://www.usgs.gov or call 1–888–ASK–USGS.

For an overview of USGS information products, including maps, imagery, and publications,
visit http://www.usgs.gov/pubprod

To order this and other USGS information products, visit http://store.usgs.gov

Suggested citation:
Heywood, C.E., 2013, Simulations of groundwater flow, transport, and age in Albuquerque, New Mexico, for a study of transport of anthropogenic and natural contaminants (TANC) to public-supply wells: U.S. Geological Survey Scientific Investigations Report 2012–5242, 51 p.

Foreword

The U.S. Geological Survey (USGS) is committed to providing the Nation with reliable scientific information that helps to enhance and protect the overall quality of life and that facilitates effective management of water, biological, energy, and mineral resources (*http://www.usgs.gov/*). Information on the Nation's water resources is critical to ensuring long-term availability of water that is safe for drinking and recreation and is suitable for industry, irrigation, and fish and wildlife. Population growth and increasing demands for water make the availability of that water, measured in terms of quantity and quality, even more essential to the long-term sustainability of our communities and ecosystems.

The USGS implemented the National Water-Quality Assessment (NAWQA) Program in 1991 to support national, regional, State, and local information needs and decisions related to water-quality management and policy (*http://water.usgs.gov/nawqa*). The NAWQA Program is designed to answer: What is the quality of our Nation's streams and groundwater? How are conditions changing over time? How do natural features and human activities affect the quality of streams and groundwater, and where are those effects most pronounced? By combining information on water chemistry, physical characteristics, stream habitat, and aquatic life, the NAWQA Program aims to provide science-based insights for current and emerging water issues and priorities. From 1991 to 2001, the NAWQA Program completed interdisciplinary assessments and established a baseline understanding of water-quality conditions in 51 of the Nation's river basins and aquifers, referred to as Study Units (*http://water.usgs.gov/nawqa/studies/study_units.html*).

In the second decade of the Program (2001–12), a major focus is on regional assessments of water-quality conditions and trends. These regional assessments are based on major river basins and principal aquifers, which encompass larger regions of the country than the Study Units. Regional assessments extend the findings in the Study Units by filling critical gaps in characterizing the quality of surface water and groundwater, and by determining water-quality status and trends at sites that have been consistently monitored for more than a decade. In addition, the regional assessments continue to build an understanding of how natural features and human activities affect water quality. Many of the regional assessments employ modeling and other scientific tools, developed on the basis of data collected at individual sites, to help extend knowledge of water quality to unmonitored, yet comparable areas within the regions. The models thereby enhance the value of our existing data and our understanding of the hydrologic system. In addition, the models are useful in evaluating various resource-management scenarios and in predicting how our actions, such as reducing or managing nonpoint and point sources of contamination, land conversion, and altering flow and (or) pumping regimes, are likely to affect water conditions within a region.

Other activities planned during the second decade include continuing national syntheses of information on pesticides, volatile organic compounds (VOCs), nutrients, trace elements, and aquatic ecology; and continuing national topical studies on the fate of agricultural chemicals, effects of urbanization on stream ecosystems, bioaccumulation of mercury in stream ecosystems, effects of nutrient enrichment on stream ecosystems, and transport of contaminants to public-supply wells.

The USGS aims to disseminate credible, timely, and relevant science information to address practical and effective water-resource management and strategies that protect and restore water quality. We hope this NAWQA publication will provide you with insights and information to meet your needs, and will foster increased citizen awareness and involvement in the protection and restoration of our Nation's waters.

The USGS recognizes that a national assessment by a single program cannot address all water-resource issues of interest. External coordination at all levels is critical for cost-effective management, regulation, and conservation of our Nation's water resources. The NAWQA Program, therefore, depends on advice and information from other agencies—Federal, State, regional, interstate, Tribal, and local—as well as nongovernmental organizations, industry, academia, and other stakeholder groups. Your assistance and suggestions are greatly appreciated.

William H. Werkheiser
USGS Associate Director for Water

Contents

Figures

Tables

Conversion Factors

SI to Inch/Pound

Multiply	By	To obtain
Length		
centimeter (cm)	0.3937	inch (in.)
meter (m)	3.281	foot (ft)
kilometer (km)	0.6214	mile (mi)
Area		
square meter (m^2)	0.0002471	acre
square kilometer (km^2)	247.1	acre
square meter (m^2)	10.76	square foot (ft^2)
square kilometer (km^2)	0.3861	square mile (mi^2)
Volume		
liter (L)	0.2642	gallon (gal)
cubic meter (m^3)	264.2	gallon (gal)
cubic meter (m^3)	35.31	cubic foot (ft^3)
cubic kilometer (km^3)	0.2399	cubic mile (mi^3)
cubic meter (m^3)	0.0008107	acre-foot (acre-ft)
Flow rate		
meter per day (m/d)	3.281	foot per day (ft/d)
cubic meter per day (m^3/d)	35.31	cubic foot per day (ft^3/d)
cubic meter per day (m^3/d)	264.2	gallon per day (gal/d)
Hydraulic conductivity		
meter per day (m/d)	3.281	foot per day (ft/d)
Hydraulic gradient		
meter per kilometer (m/km)	5.27983	foot per mile (ft/mi)
Transmissivity*		
meter squared per day (m^2/d)	10.76	foot squared per day (ft^2/d)

Temperature in degrees Celsius (°C) may be converted to degrees Fahrenheit (°F) as follows:

°F=(1.8×°C)+32

Temperature in degrees Fahrenheit (°F) may be converted to degrees Celsius (°C) as follows:

°C=(°F–32)/1.8

Vertical coordinate information is referenced to the North American Vertical Datum of 1988 (NAVD 88)

Horizontal coordinate information is referenced to the North American Datum of 1983 (NAD 83).

Altitude, as used in this report, refers to distance above the vertical datum.

Concentrations of chemical constituents in water are given either in milligrams per liter (mg/L) or micrograms per liter (µg/L).

Acronyms and Abbreviations

ABCWUA	Albuquerque Bernalillo County Water Utility Authority
BFH	Boundary flow and head package
CFC	Chlorofluorocarbon
DEM	Digital Elevation Model
DNAPL	Dense Non-Aqueous Phase Liquid
GIS	Geographic Information System
HOB	Head Observation Package
LGR	Local Grid Refinement
LPF	Layer-Property Flow package
MNW2	Multi-node well package
MRGB	Middle Rio Grande Basin
MTBE	Methyl *tert*-butyl ether
NAWQA	National Water-Quality Assessment
PCE	Tetrachloroethene
pmc	Percent modern carbon
ppt	Parts per trillion
pptv	Parts per trillion by volume
PSW	Public-Supply Well
RCH	Recharge package
RMSE	Root mean square error
SSW	Studied Supply Well
TANC	Transport of Anthropogenic and Natural Contaminants
TCE	Trichloroethene
TDS	Total Dissolved Solids
TOB	Transport Observation Package
TPROGS	Transition Probability Geostatistical Software
TU	Tritium units
TVD	Total Variation Diminishing scheme
USGS	U.S. Geological Survey
VOC	Volatile Organic Compound

Simulations of Groundwater Flow, Transport, and Age in Albuquerque, New Mexico, for a Study of Transport of Anthropogenic and Natural Contaminants (TANC) to Public-Supply Wells

By Charles E. Heywood

Abstract

Vulnerability to contamination from manmade and natural sources can be characterized by the groundwater-age distribution measured in a supply well and the associated implications for the source depths of the withdrawn water. Coupled groundwater flow and transport models were developed to simulate the transport of the geochemical age-tracers carbon-14, tritium, and three chlorofluorocarbon species to public-supply wells in Albuquerque, New Mexico. A separate, regional-scale simulation of transport of carbon-14 that used the flow-field computed by a previously documented regional groundwater flow model was calibrated and used to specify the initial concentrations of carbon-14 in the local-scale transport model. Observations of the concentrations of each of the five chemical species, in addition to water-level observations and measurements of intra-borehole flow within a public-supply well, were used to calibrate parameters of the local-scale groundwater flow and transport models.

The calibrated groundwater flow model simulates the mixing of "young" groundwater, which entered the groundwater flow system after 1950 as recharge at the water table, with older resident groundwater that is more likely associated with natural contaminants. Complexity of the aquifer system in the zone of transport between the water table and public-supply well screens was simulated with a geostatistically generated stratigraphic realization based upon observed lithologic transitions at borehole control locations. Because effective porosity was simulated as spatially uniform, the simulated age tracers are more efficiently transported through the portions of the simulated aquifer with relatively higher simulated hydraulic conductivity. Non-pumping groundwater wells with long screens that connect aquifer intervals having different hydraulic heads can provide alternate pathways for contaminant transport that are faster than the advective transport through the aquifer material. Simulation of flow and transport through these wells requires time discretization that adequately represents periods of pumping and non-pumping. The effects of intra-borehole flow are not fully represented in the simulation because it employs seasonal stress periods, which are longer than periods of pumping and non-pumping. Further simulations utilizing daily pumpage data and model stress periods may help quantify the relative effects of intra-borehole versus advective aquifer flow on the transport of contaminants near the public-supply wells. The fraction of young water withdrawn from the studied supply well varies with simulated pumping rates due to changes in the relative contributions to flow from different aquifer intervals.

The advective transport of dissolved solutes from a known contaminant source to the public-supply wells was simulated by using particle-tracking. Because of the transient groundwater flow field, scenarios with alternative contaminant release times result in different simulated-particle fates, most of which are withdrawn from the aquifer at wells that are between the source and the studied supply well. The relatively small effective porosity required to simulate advective transport from the simulated contaminant source to the studied supply well is representative of a preferential pathway and not the predominant aquifer effective porosity that was estimated by the calibration of the model to observed chemical-tracer concentrations.

Introduction

Clean groundwater is a valuable source of public drinking water throughout the world, especially in arid areas with limited surface-water availability. Groundwater wells that supply drinking water to the public are vulnerable to contamination from natural and anthropogenic (manmade) sources. An improved understanding of the sources of water to public-supply wells and the processes that lead to contamination is needed to reduce the incidence of natural and manmade contamination in wells that supply this resource. In order to increase understanding of the factors that affect transport of anthropogenic and natural contaminants (TANC) to public-supply wells, the U.S. Geological Survey (USGS) National Water-Quality Assessment (NAWQA) Program has conducted a series of groundwater studies in areas representative of the different types of aquifer systems that exist in the United States (Paschke, 2007). One component of these studies has been the use of groundwater models to investigate the sources of contaminants, their fate and transport within groundwater systems, and their effects on public-supply-well water quality at regional and local scales. The regional-scale groundwater flow models helped to establish relations between hydrogeologic and land-use variables with observed public-supply-well water-quality trends. Local-scale groundwater flow and transport models nested within the regional-scale model areas have provided increased detail for understanding physical and chemical processes that affect public-supply-well water quality.

Supply-well vulnerability can be characterized by the groundwater age, which is a measure of the travel time from recharge-source areas, and associated implications for the depths of the withdrawn water. In general, young groundwater in a well can indicate a vulnerability to manmade contaminants in the aquifer. However, management actions based on avoidance of young groundwater can lead to increased vulnerability to natural contaminants, such as arsenic (Hinkle and others, 2009). Because the age of groundwater is strongly correlated to its depth within an aquifer system, the groundwater-age distribution measured in a supply well can be used to imply the source depths of the withdrawn water and characterize its vulnerability to contamination from either deep or shallow sources. Although natural contaminants may be associated with sources that reside in either shallow or deep portions of an aquifer system, anthropogenic contamination usually originates from a shallow source.

Within the Middle Rio Grande Basin (MRGB) in New Mexico (fig. 1) considered in this study, aquifer-system recharge is thought to principally occur from losing reaches of the Rio Grande (Plummer and others, 2004) and adjacent to mountain fronts, where overland runoff across relatively impermeable bedrock first infiltrates into more permeable basin-fill sediments. Infiltration of precipitation through the thick unsaturated zone (tens to hundreds of meters) that exists over most of the remainder of the basin is not thought to substantively occur or to contribute to recharge away from the mountain fronts. Long times are required for groundwater to travel from either of these recharge locations; therefore, anthropogenic contaminants, such as volatile organic compounds (VOCs) and the groundwater age-tracers tritium and chlorofluorocarbons (CFCs), which are indicative of a component of relatively recent recharge, would not be present at shallow and intermediate depths within the study area if these distant recharge locations were the only pathways for young water to enter the groundwater system. However, measureable concentrations of these chemicals from production and observation wells within the study area are documented in the companion report by Bexfield and others (2012), implying the existence of closer sources of young water and relatively fast pathways for transport of that water into wells.

For the TANC study area of the MRGB, a regional-scale groundwater model simulated transient groundwater flow through 2008 and was used to compute areas contributing recharge and the transport times from recharge areas to public-supply wells (Bexfield and others, 2011). Although the discretization of this regional-scale model was finer than that used for previous models of the MRGB, it remained too coarse to adequately simulate heterogeneity within the shallow and intermediate depths of the aquifer system and pathways through which anthropogenic contaminants and young groundwater-age tracers may be transported to wells screened at such depths within the aquifer system.

The local-scale groundwater model documented in this report includes additional sources of local recharge that can simulate introduction of young (post-1950s) recharge water, and its entrained age-tracers or contaminants, into the saturated aquifer system. The simulated mixing of this younger water with resident older groundwater is constrained by observations of the concentrations of carbon-14, tritium, and CFCs. Contaminants associated with the younger water may be transported relatively quickly through the more hydraulically conductive portions of the simulated aquifer system. Well boreholes connecting aquifer intervals with different hydraulic heads can provide an alternate pathway for groundwater to flow and transport dissolved solutes.

The transient groundwater flow model simulates the effects of seasonal changes in pumping rates on hydraulic gradients and consequent transport rates between different aquifer intervals and the screened intervals of the simulated withdrawal wells. Known sources of anthropogenic contamination, including VOCs, exist within the local study area, and contaminants believed to have originated at those sources have been observed at various observation wells (Bexfield and others, 2012). The likely advective transport pathways and contaminant source-release times were investigated with the MODPATH particle-tracking program.

Purpose and Scope

This report describes the construction and calibration of a local-scale model that simulates the groundwater flow, transport, and age within the local-scale TANC study area in Albuquerque, New Mexico. The local-scale model was numerically coupled with a new local-grid-refinement technique to a previously documented (Bexfield and others, 2011) regional-scale groundwater model of the MRGB of New Mexico. This coupling provided accurate boundary conditions around the perimeter of the local-scale model throughout the course of the 108-year simulation, during which the simulated water levels and flows across model boundaries changed significantly. The simulated groundwater flow field computed by the local-scale model was used as input for an associated solute-transport model, which encompasses the same study area. The simulated tracers of young groundwater (^3H and CFCs) were surrogates for potentially harmful anthropogenic contaminants

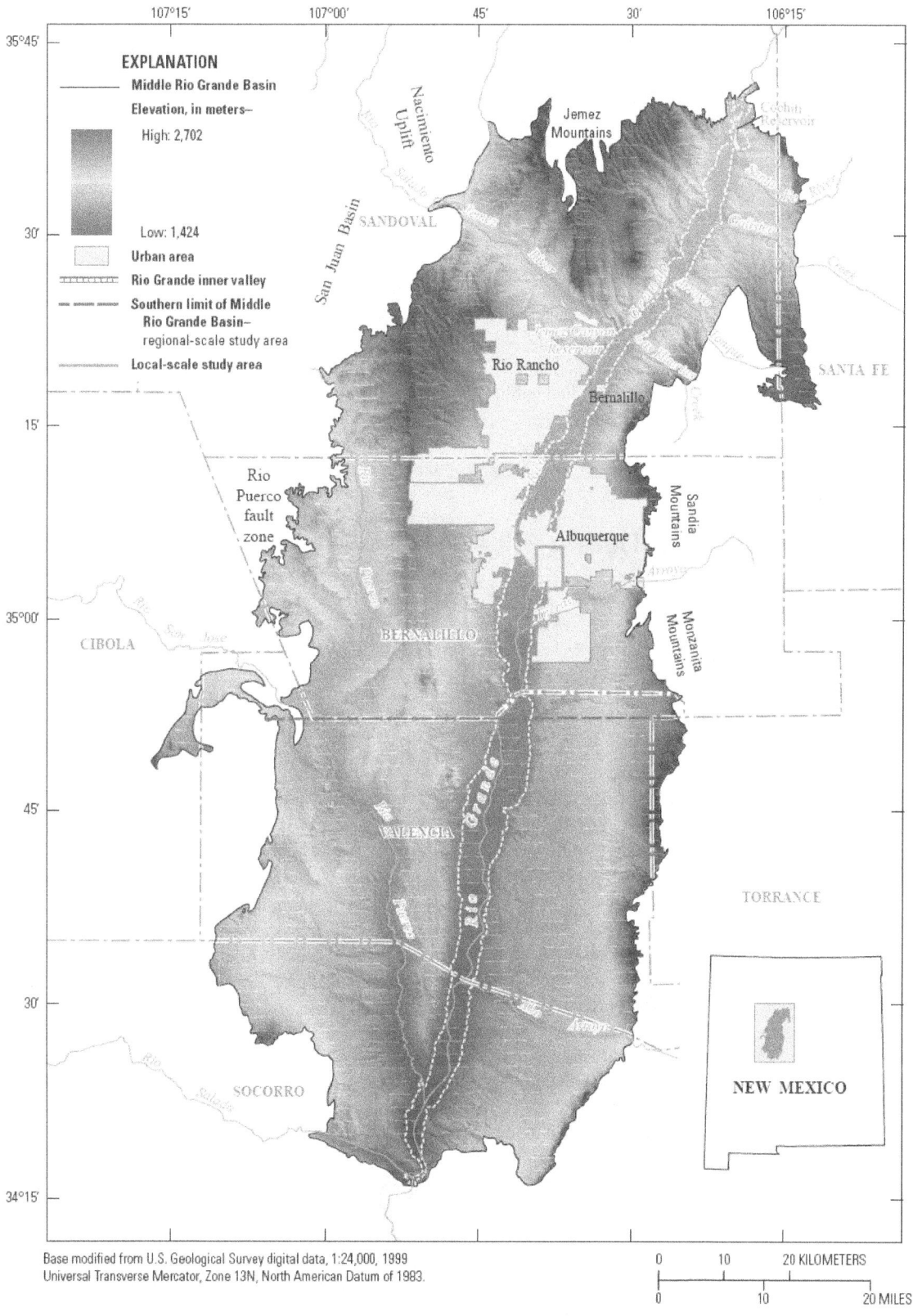

Figure 1. Shaded relief of the Middle Rio Grande Basin in New Mexico, showing major regional features and the location of the local-scale groundwater model.

(such as VOCs) that may be transported along similar pathways in the aquifer system. Simulation of carbon-14 concentrations required specification of initial concentration conditions that realistically represented the carbon-14 concentration distribution throughout the three-dimensional local-scale-model domain prior to the onset of groundwater pumping and recharge of water with carbon-14 concentrations affected by nuclear-bomb testing. When combined with observed water-level and borehole-flow observation data, observed isotopic and chemical concentration data provided an observation database for calibration of groundwater flow and solute-transport model parameters with non-linear regression.

A companion report by Bexfield and others (2012) describes the hydrogeology, water chemistry, and factors affecting the transport of selected anthropogenic and natural contaminants in the zone of contribution to the studied public-supply well in the local-scale TANC study area. Together, these reports enable further future comparisons amongst TANC local-scale study areas.

Description of Local-Scale Model Area

Bexfield and others (2012) identified an area for detailed study within metropolitan Albuquerque, the largest city in the MRGB in New Mexico (fig. 1). The local-scale model described in this report was designed to simulate groundwater flow and solute transport beneath this area. The horizontal area encompassed by the model is 24 square kilometers (km^2) [4 kilometers (km) in the east-west direction and 6 km in the north-south direction] and is located approximately south of Lomas Boulevard, north of the Albuquerque International Airport, east of the Rio Grande, and west of Girard Boulevard (fig. 2). This area encompasses some of the inner valley east of the Rio Grande where agriculture began in the 1900s and where areas of residential, commercial, and industrial land use were developed in the 1930s. Current land use within this area can be characterized as "urban," with several recreational parks and small open-space areas.

The western half of the local-scale study area lies within the Rio Grande inner valley and is relatively flat with altitudes ranging from 1,505 to 1,525 meters (m). An abrupt change in slope separates the inner valley from the terrace which gradually rises toward the mountain front to the east. The eastern half of the local-scale study area is on this terrace with moderate relief and altitudes that range from about 1,560 to 1,620 m.

The local-model area (fig. 2) contains at least seven public water-supply wells (PSWs), one of which Bexfield and others (2012) selected for intensive study, known hereafter as the studied supply well (SSW). Three known sites of groundwater contaminated with chlorinated solvents, including trichloroethene (TCE), are present in the local-model area, and petroleum storage tanks have leaked at two or more sites.

The SSW is a high-production public-supply well located in an area where the depth to the water table is greater than 30 m. Despite having well screens installed at an even greater depth, anthropogenic contamination in the forms of chlorinated solvents and methyl *tert*-butyl ether (MTBE) has been measured at low concentrations in the SSW. Chlorinated solvents and MTBE have also been detected at concentrations far below drinking-water standards in other nearby PSWs, which were installed in the 1960s and 1970s (Bexfield and others, 2012).

Of the three sites known to be contaminated with chlorinated solvents, the site located 3 km northwest of the SSW is of particular interest as a likely source of contaminants to the SSW. The documented contamination at this site, where a (now defunct) dry-cleaning facility was located, is in an area of the Rio Grande inner valley where the depth to the water table is 9 m or less. Monitoring wells have indicated the presence of younger, more contaminated water beneath older, less contaminated ground water (U.S. Environmental Protection Agency, 2011). It is likely that vertical contaminant transport to more than 60 m below the water table has resulted from pumping of supply wells that induced downward hydraulic gradients and (or) from flow down well boreholes between the contamination site and the SSW.

A network of monitoring wells provided ground-water level and chemical data within the local study area. In addition to the 12 existing wells that were utilized for data collection, 13 new wells at 4 flow-path (FP) locations were installed for the TANC study (Bexfield and others, 2012). Wellbore flow and water-chemistry data were also obtained from various depths within the SSW.

Previous Investigations

Summaries of the extensive previous hydrogeologic investigations are contained in publications by Thorn and others (1993), Bartolino and Cole (2002), Bexfield (2010), and Bexfield and others (2011, 2012). Because the present work is based in part on previous investigations involving groundwater modeling, a brief synopsis of the recent pertinent studies is included here. McAda and Barroll (2002) developed a transient regional groundwater flow model of the Albuquerque basin that simulated predevelopment conditions and anthropogenic changes for the years 1900 through March 2000. Sanford and others (2004) used carbon-14 concentrations and hydrochemical-zone locations to constrain the hydraulic parameters of an alternative regional groundwater flow model that was calibrated with the program UCODE (Poeter and Hill, 1998). Bexfield and others (2011) modified the McAda and Barroll (2002) model by refining the horizontal discretization, adding transient stress data (representing periods through Dec. 31, 2008), simulating urban recharge, and revising the hydraulic parameters, which were estimated by calibrating the model using the program PEST (Doherty, 2004). The basin-fill thickness estimates, refined on the basis of recent analyses of seismic reflection data and isostatic-residual gravity anomalies (Grauch and Connell, 2013), were incorporated into the model-layer thicknesses of the version of the regional model documented and utilized for the analyses in this report.

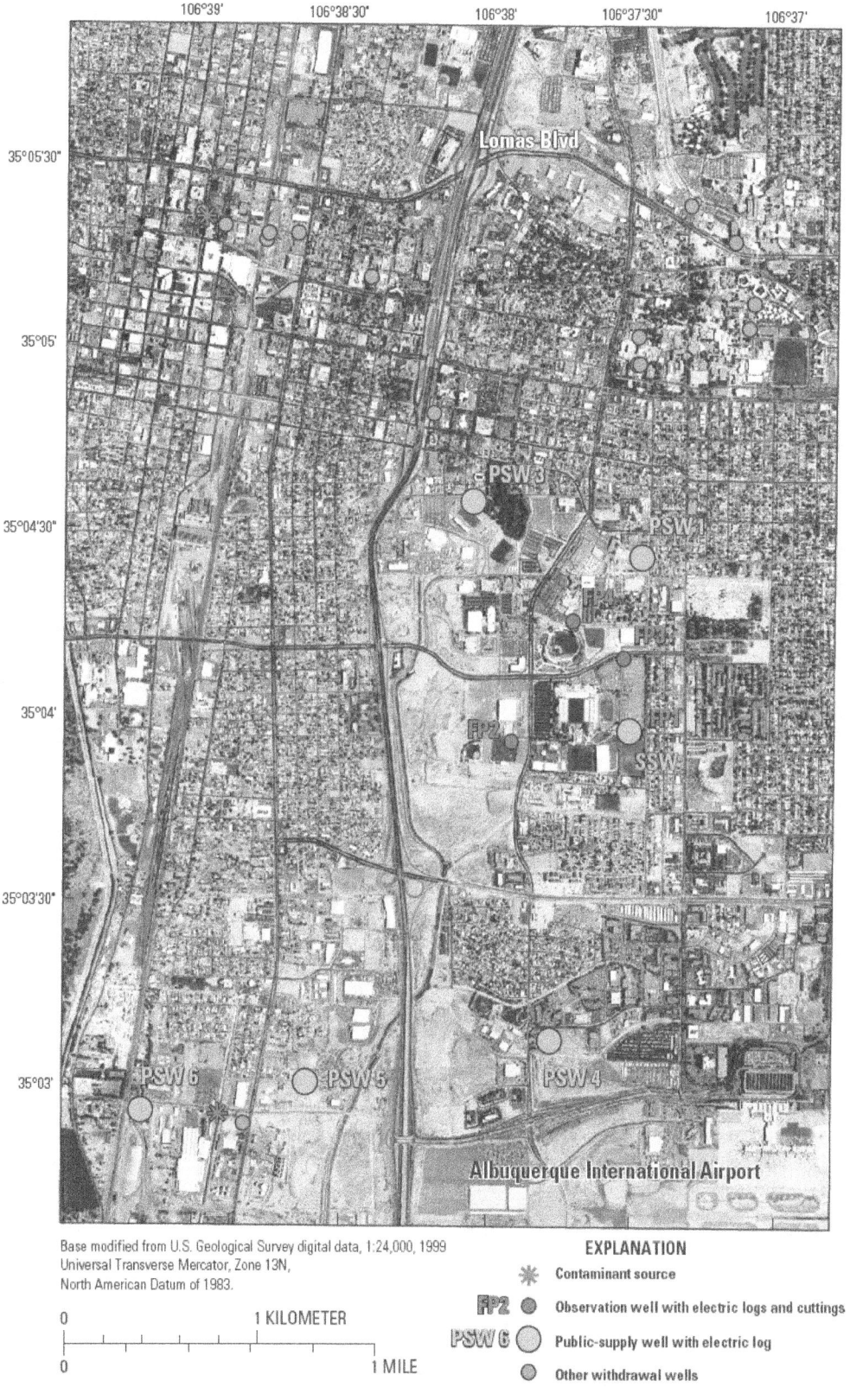

Figure 2. Land-surface imagery of the local-scale model area from 2005, and locations of public-supply wells and selected known contaminant sources, Albuquerque, New Mexico.

Hydrogeology of the Local-Scale Model Area

The alluvial-fill within the MRGB consists primarily of the Santa Fe Group of late Oligocene to middle Pleistocene age, which consists of unconsolidated to moderately consolidated sediments deposited in fluvial, lacustrine, or piedmont-slope environments. The screened intervals of most wells within the study area are within the Rio Puerco member of the Ceja Formation of the Santa Fe Group, which consists of fluvial deposits (sand, gravel, and mud) of Pliocene age (Connell and others, 1998; Connell, 2006). Within the local-scale model area, the Ceja Formation is overlain by Quaternary alluvium within the Rio Grande inner valley and overlain by the Sierra Ladrones Formation of Pliocene to lower Pleistocene age beneath the higher terrace areas to the east (Connell, 2006). The lower Atrisco member of the Ceja Formation is generally finer grained and lies beneath the screened intervals of most wells in the study area.

The natural, predevelopment direction of groundwater flow through this area was from the north-northeast and toward the Rio Grande. Considerable groundwater withdrawals in the area have redirected the flow directions to the east and steepened both horizontal and vertical head gradients. More detailed discussions of the climate, hydrology, geology, land use, topography, water chemistry, installation of flow-path (FP) well nests, and other aspects of the regional and local-scale study area are found in the companion reports Bexfield and others (2011, 2012).

Aquifer-Pumping Test

Operation of the SSW in February 2008 provided data that were used to obtain hydraulic conductivity estimates of the sediments in which the SSW was screened. Following a 2-month period of inactivity, the SSW was pumped for about 10 hours and was subsequently turned off for several days. The time-drawdown data recorded in the FP1 nest of wells located about 33 m east of the SSW were used to calculate a transmissivity of the aquifer in the screened interval of the SSW. The analysis employed a Neuman (1974) unconfined solution assuming partial penetration of the SSW to the top of the relatively fine-grained Atrisco member of the Ceja Formation (Scott Christenson, U.S. Geological Survey (retired), written commun., 2011). Dividing the analytically derived transmissivity values for each of the four piezometers in the FP1 nest by the length of the screened interval of the SSW yielded four estimates of hydraulic conductivity in this interval, which range from 3.1 to 7.9 meters per day (m/d). These values and the average of the four interpreted values (4.5 m/d) are similar to previous estimates of about 6.1 m/d based on aquifer tests that used public-supply wells in the vicinity of the SSW (Thorn and others, 1993).

Characterization of Tertiary Alluvium

Within the local study area, electric-resistivity and lithologic logs from 10 sites and borehole cuttings from 4 sites were used to characterize lithologic variations with depth (fig. 2). In addition, four resistivity logs of the Burton Well-Field Wells (#1–4), located within 1.2 km east of Girard Blvd., were used for lithologic characterization. Short-normal resistivity logs from the 14 sites, spanning resistivity ranges on the order of 100 ohm-meters (ohm-m), provided the primary basis for lithologic interpretation. The portions of the individual resistivity logs used for classification ranged in length from 47 to 134 m; the average thickness was 108 m. Major resistivity-response patterns correlated well between boreholes in the north-south direction and weakly in the east-west direction. The enhanced correlation in the north-south direction may be related to the primary flow direction of the fluvial depositional system, which was predominantly north-south.

A 3-m depth interval was chosen for characterizing lithology in each borehole, which was somewhat greater than the resolution of the short-normal resistivity logs, and this length was for the vertical finite-difference cell dimension within the finest-resolution depth interval of the numerical flow model. Most of the driller's lithologic logs summarized lithology at intervals much larger than 3 m and were therefore unsuitable for classifying lithology at 3-m intervals. Driller's logs from the four FP monitoring wells were primarily summarized at 3-m intervals and were based on drilling rate and subjective judgment of drilling vibration, in addition to summaries of borehole cuttings collected from each 3-m depth interval. The cuttings were collected at the top of the borehole with a fine strainer that allowed capture of material that was approximately sand sized and larger. Additional samples of sand-sized cuttings were collected from the drilling-mud recycling equipment. Because drilling mud (consisting of clay- and silt-sized particles) was used to excavate cuttings from the borehole, the quantity of clay- and silt-sized cuttings could not be reliably determined. Other complications associated with use of the cuttings to represent the lithology of the borehole include (1) the small sample taken from each 3-m borehole interval, (2) increased travel distance to the surficial collection point with increased borehole depth and the consequent greater uncertainty of the representative depth of cuttings, and (3) greater upward travel-time for gravel-sized cuttings relative to sand-sized cuttings.

The alluvial aquifer in the local-study area is dominantly sand fill with gravel lenses and thin layers of finer-grained sediment. In order to correlate aquifer lithology with the response of the resistivity logs, two lithologic logs from borehole FP1, one prepared onsite by the drillers and a second prepared in a laboratory after washing of the cuttings to remove residual drilling mud (table 1), were compared with the resistivity log from borehole FP1 (fig. 3). The resistivity logs show distinct patterns within three particular depth ranges: large-amplitude fluctuations between high- and low-resistivity peaks from

Table 1. Drillers' and laboratory lithologic logs of borehole FP1 in Albuquerque, New Mexico, 2007.

Depth (meters below land surface)	Drillers' lithologic log	Lithologic log based on laboratory analysis of borehole cuttings[1]
55–58	Sand, small gravel	Fine silty sand
58–61	Sand, small gravel, silty clay	*NS*[2]
61–64	Sand, silty clay	Fine sand
64–67	Sand, silty clay, small gravel	Coarse to medium sand
67–70	Silty clay, sand	Coarse to medium sand
70–73	Sand, small gravel	Coarse to medium sand
73–76	Sand, small gravel	Very fine to fine sand
76–79	Sand, small gravel	Very fine to fine sand
79–82	Sand, silty clay	Medium to coarse sand
82–85	Core	Medium to coarse sand
85–88	Core	Medium to coarse sand
88–91	Sand, silty clay	Very fine to fine sand
91–94	Sand, small gravel	Coarse to very coarse sand
94–98	Sand, gravel, silty clay	Very fine to fine sand
98–101	Small gravel, sand	Coarse to very coarse sand
101–104	Small gravel, sand	Fine sand
104–107	Sand, silty clay	Fine sand
107–110	Sand, silty clay	Fine sand
110–113	Sand, silty clay	Very fine to fine sand
113–116	Sand, small gravel, silty clay	Very fine to fine sand
116–119	Sand, sandy clay	Very fine to fine sand
119–122	Sand, sandy clay	Very fine to fine sand
122–125	Sand, silty clay	Very fine to fine sand
125–128	Sand, silty clay	Very fine to fine sand
128–131	Sand, silty clay	Very fine to fine sand
131–134	Sand, silty clay, small gravel	Very fine silty sand
134–137	Sand, silty clay	Very fine silty sand
137–140	Sand, silty clay	Very fine silty sand
140–143	Sand, silty clay	Coarse sand with very fine silty sand
143–146	Sand, silty clay, small gravel	Coarse sand with very fine silty sand
146–149	Sand, silty clay, small gravel	Coarse sand with very fine silty sand
149–152	Core	Very fine clay-rich sand
152–155	Core	Very fine clay-rich sand
155–158	Sand, gravel	Very fine clay-rich sand
158–162	Gravel	Very fine clay-rich sand
162–165	Gravel	Very fine silty sand
165–168	Sand, small gravel	Very fine silty sand
168–171	Sand, small gravel	Coarse sand

[1] Silt and clay were generally not included in the descriptions because they were difficult to distinguish from the drilling mud.

[2] *NS* - No sample.

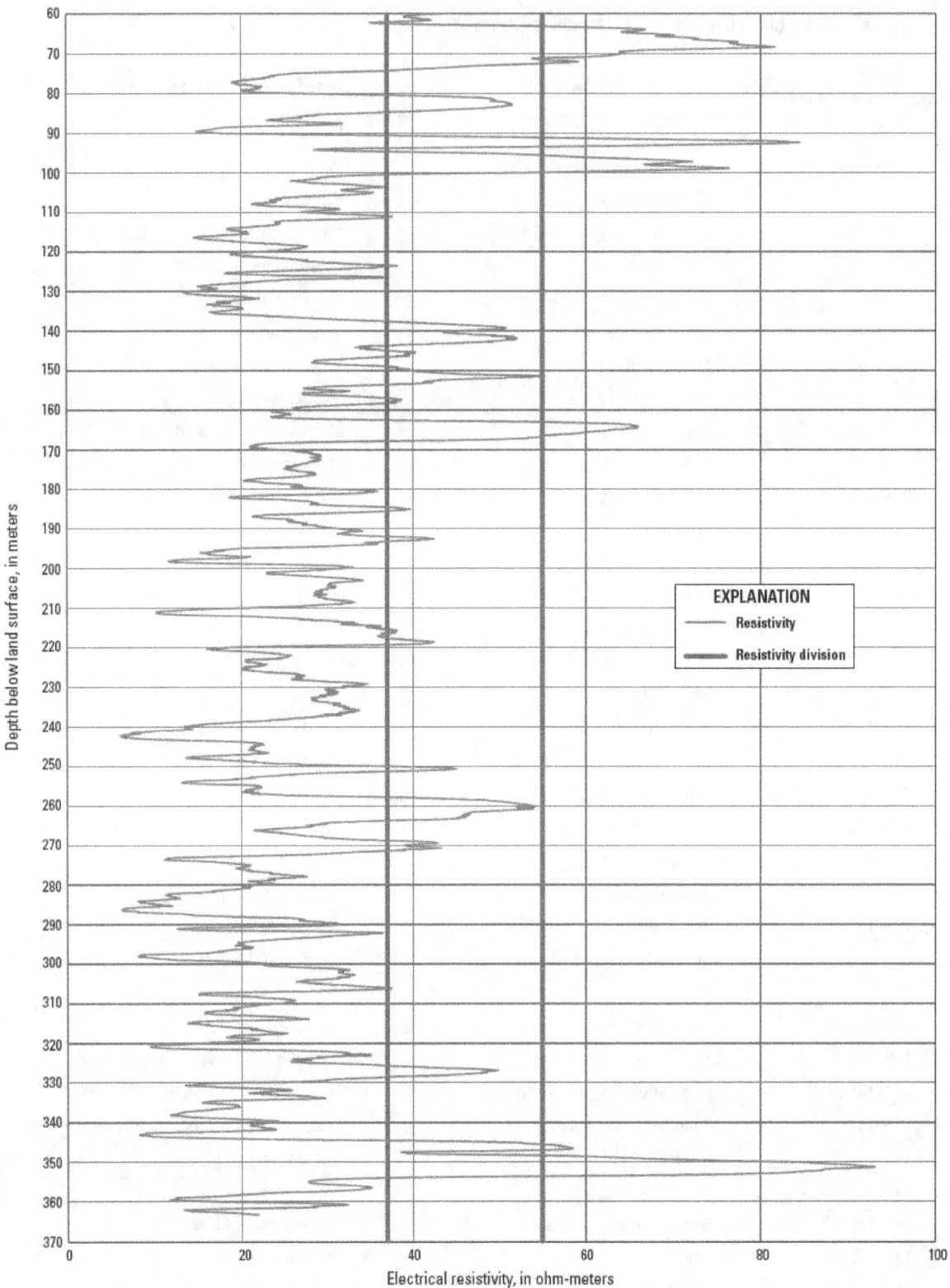

Figure 3.   Short-normal resistivity log of borehole FP1 in Albuquerque, New Mexico, 2007.

depths of 60 to 100 m, small-amplitude fluctuations and low resistivity between depths of 100 and 137 m, and intermediate-amplitude fluctuations around moderate resistivity values between depths of 137 and 171 m. Comparison of the resistivity and lithologic logs indicates that between depths of 60 and 100 m, the high resistivity peaks are generally associated with gravels and coarse sands, while the low peaks are generally associated with fine sands and (or) silty clays. Between depths of 100 and 137 m, the low resistivity is again generally associated with fine sands and silty clays. Between depths of 137 and 171 m, gravels, sands, and silty clays exist, but the correlation between the lithologic and resistivity logs is poor. This may be caused by less-representative cuttings of the actual lithology with increased depth of the borehole.

The resistivity log for borehole FP1 was divided into three resistivity intervals based on the sediment lithology and the electrical resistivity values. The division between the low and intermediate interval, at 37 ohm-m, is the high-resistivity limit of measurements between 100 and 137 m depth. The division between the intermediate and upper interval, at 55 ohm-m, is the low-resistivity limit of measurements between 55 and 100 m in which primarily gravel or coarse sand lithologies exist. Three-meter depth intervals where measured resistivity was generally less than 37 ohm-m were assumed to be fine sand and silty clays. Three-meter depth intervals where measured resistivity was generally larger than 55 ohm-m were assumed to represent gravel and coarse sand. Three-meter depth intervals where measured resistivity between these two values were assumed to represent a mixture of sands with lenses of gravel and silty clay.

Because electrical-resistivity data from the 14 boreholes were measured using different resistivity sensors in various holes, the borehole resistivity logs do not all provide the same response to materials of identical lithology. As a result, absolute resistivity values could not be compared among logs from the different boreholes. However, because the sediments sampled during drilling of each of the 14 boreholes were deposited by the same ancestral fluvial system, it is assumed that the patterns in the relative resistivity response in each borehole correspond to similar changes in the subsurface lithology. Under this assumption, lower and upper divisions were plotted on the resistivity logs of the 13 other boreholes by either (1) correlation with the resistivity log of borehole FP1 or (2) observing patterns of low, medium, and high resistivity responses similar to the resistivity log from borehole FP1. For each resistivity log, the upper resistivity division varied but was usually 20–40 ohm-m higher than the lower division.

Based on the observed correlation between electric-log responses and grain sizes in the sample cuttings, as well as documented relations among permeability, grain size, and electric-log responses (Alger, 1966; Croft, 1971) and between hydraulic conductivity and grain size (Hazen, 1911; Shepherd, 1989), the gross electric-log responses were assumed to indicate grain size and relative hydraulic conductivity. Aquifer depth intervals corresponding to low, intermediate, and high resistivity intervals, such as those previously described for

well FP1, were assigned to one of three hydrofacies representing fine, moderate, and coarse grain-sized zones, which, in turn, have relatively low, intermediate, and high hydraulic conductivity, respectively. The simulated hydraulic conductivities for each of these three zones were estimated by the parameter-estimation process described in the Model Calibration section of this report.

The distribution of the low, intermediate, and high resistivity intervals in the 14 boreholes were used to synthesize 10 realizations of hydrostratigraphy using Transition Probability Geostatistical Software (TPROGS) (Carle, 1999). Corresponding fine, moderate, and coarse hydrofacies occupied 59, 33, and 8 percent, respectively, of the total volume sampled. Each realization honored the stratigraphic data at the borehole control sites and had identical transition probabilities, mean hydrofacies lengths, and other geostatistical properties. The synthetic hydrostratigraphic realizations resemble maps of similar surficial alluvial deposits in the Albuquerque basin deposits in scale, orientation, and relative distribution and shapes of the hydrofacies.

Each realization was generated with a base altitude of 1,341 m and discretized the hydrofacies on a grid of 40 x 60 x 60 cells in the east-west, north-south, and vertical directions, respectively, with cell spacing of 100 m in both horizontal directions and 3 m vertically. The portion of each realization corresponding to thirty-two 3-m-thick local-scale model layers (realization layers 14 through 45) were extracted to simulate a complex hydraulic conductivity distribution in the portion of the local groundwater flow model between the altitudes of 1,381 and 1,478 m. Block and dissected perspectives (fig. 4) of a hydrofacies realization depict the simulated continuity amongst the three facies.

Methods for Simulation of Groundwater Flow and Solute Concentrations

Although this study focuses on simulation of groundwater flow and solute transport within the local study area, simulations of regional groundwater flow and carbon-14 transport were necessary to provide boundary and initial conditions for the local-scale flow and transport simulations. The boundary and initial conditions required for the local-scale groundwater flow model were provided by use of numerical coupling with a modified version of the regional-scale groundwater flow model documented by Bexfield and others (2011).

Coupling of Regional and Local-Scale Groundwater Flow Models

The lateral boundaries of the regional-scale model correspond with physical hydrologic or geologic features, but those of the local-scale model do not; consequently, groundwater

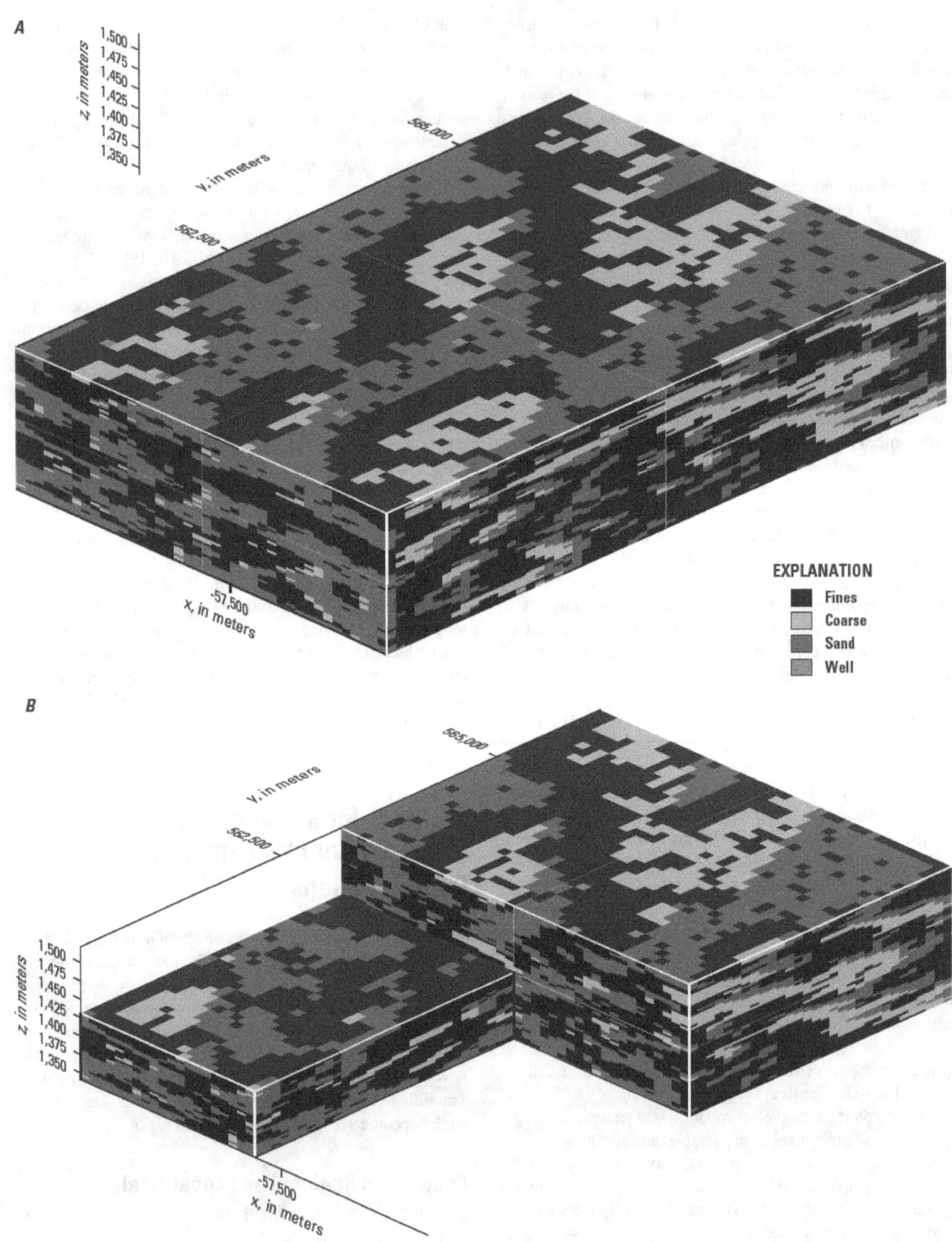

Figure 4.   (*A*) Block and (*B*) dissected perspective views of one hydrofacies realization in the local model area.

levels and flows across lateral boundaries of the local-scale model change over time due to groundwater pumpage and other transient conditions that are simulated by both the regional and local-scale models. In order to provide physically realistic lateral boundary conditions for the local-scale model that are based on the regional groundwater system, the transient groundwater levels on the local-scale-model lateral boundaries were computed with the local grid refinement (LGR) method.

MODFLOW-LGR (Mehl and Hill, 2013) was used to simulate constant-density groundwater flow at the regional and local-area scales. MODFLOW-LGR is a version of the MOD-FLOW-2005 code (Harbaugh, 2005) that enables numerical coupling of a "parent model," which is typically regional in scale, with a "child model" that has finer-detailed spatial discretization in an area where more accurate simulation is needed. The local-scale, or "child model," developed for this study was coupled to a modified version of the regional groundwater flow simulation documented by Bexfield and others (2011). Initially, the local-scale model grid was designed to couple with the "shared node" method (LGR1) of local-grid refinement (Mehl and Hill, 2005). While this coupling method was satisfactory for a steady-state simulation and several early transient stress periods, the method failed during subsequent transient stress periods because the simulated water-table decline caused "drying out" and deactivation of finite-difference cells along the parent–child model interface that are required for numerical coupling. Fortunately, an alternative code, LGR2 (Mehl and Hill, 2013), became available, which uses a new "ghost node" method of local grid refinement (Dickinson and others, 2007). Testing with a beta-version of LGR2 was eventually successful in coupling the local- to regional-scale models during all transient stress periods and was subsequently used for all groundwater flow simulations.

In flow simulations with the coupled regional-scale and local-scale flow models, regional-flow computations within the area encompassed by the local-model (fig. 2) are de-activated in the regional model. When using LGR2, heads within each model and fluxes across the interface between the two models are iteratively computed for each time step by both regional (parent) and local-scale (child) models until satisfactory convergence between the values computed by both models is achieved. This numerical coupling effectively specifies the transient heads around the perimeter of the local-scale model.

After model calibration produced a finalized set of flow-model parameters, the calibrated local-scale flow simulation could be run independently of the regional-scale model using the Boundary Flow and Head Package (BFH2) of MOD-FLOW-LGR (Mehl and Hill, 2013) and a previously saved file containing specified-head boundary conditions along the perimeter of the local-scale model. Running the decoupled local-scale model is faster and convenient if the model must be rerun to produce the head and cell-by-cell flow files

required for subsequent particle tracking and solute-transport simulations.

Specification of initial carbon-14 concentrations was required for the local-scale solute-transport model. Although it could be advantageous to couple a local-scale transport simulation to the regional-scale transport simulation documented in appendix 1 of this report, a local-grid-refinement method for solute transport (analogous to MODFLOW-LGR) is not currently available. Accordingly, the initial carbon-14 concentrations required for the local-scale transport model were interpolated from concentrations simulated for the steady-state stress period of the regional-scale carbon-14 transport simulation documented in appendix 1.

Modification of the Regional Groundwater Flow Model

Recent reprocessing and subsequent reinterpretation of seismic data combined with modeling of gravitational anomalies have provided a revised interpretation of the spatial distribution of thickness of alluvial-fill sediments in the MRGB (Grauch and Connell, in press). The thickness of finite-difference layers in the model previously documented in Bexfield and others (2011) was revised somewhat to simulate this new interpretation of the depth of alluvial-fill sediments. As depicted in figure 5, the ranges of steady-state saturated thickness of layers 2, 3, 4, and 5 are 15–23, 30–47, 65–103, and 24–184 m, respectively. The base of layer 5 is at an altitude 244 m below the Rio Grande, and it maintains that altitude perpendicular to the trend of the river except where basement rock is at a higher altitude near the basin perimeter. Layers 6 and 7 have constant thicknesses of 183 and 305 m, respectively, except near the basin perimeter, where layer-bottom altitudes rise with the basement rock and the thickness of these layers thins to about 30 meters. The top of layer 8 is at an altitude 732 m below the Rio Grande, except near the basin perimeter, where the layer-bottom altitude also rises, causing a range in layer thickness from 15 to 1,532 m. The thickness of layer 9 ranges from 32 to 3,064 m. Cells in layers 1–9 are active where the base of the cell is higher than the base of the Santa Fe Group basin fill. Because the greatest thickness changes occur in the deepest layer (regional model layer 9), within which the simulated hydraulic conductivity is small, the changes in simulated regional groundwater flow are relatively minor.

The temporal discretization of the regional model was refined by replacing a stress period representing the autumn of 2007 with three stress periods, as described in more detail in the following section. This modification was required to enable coupling with the local-scale model, which must have identical time discretization.

When the revised regional-scale model is coupled to the embedded local-scale model with LGR2, the portion of the regional model domain encompassed by the local-scale model is deactivated, and that portion of the model domain is

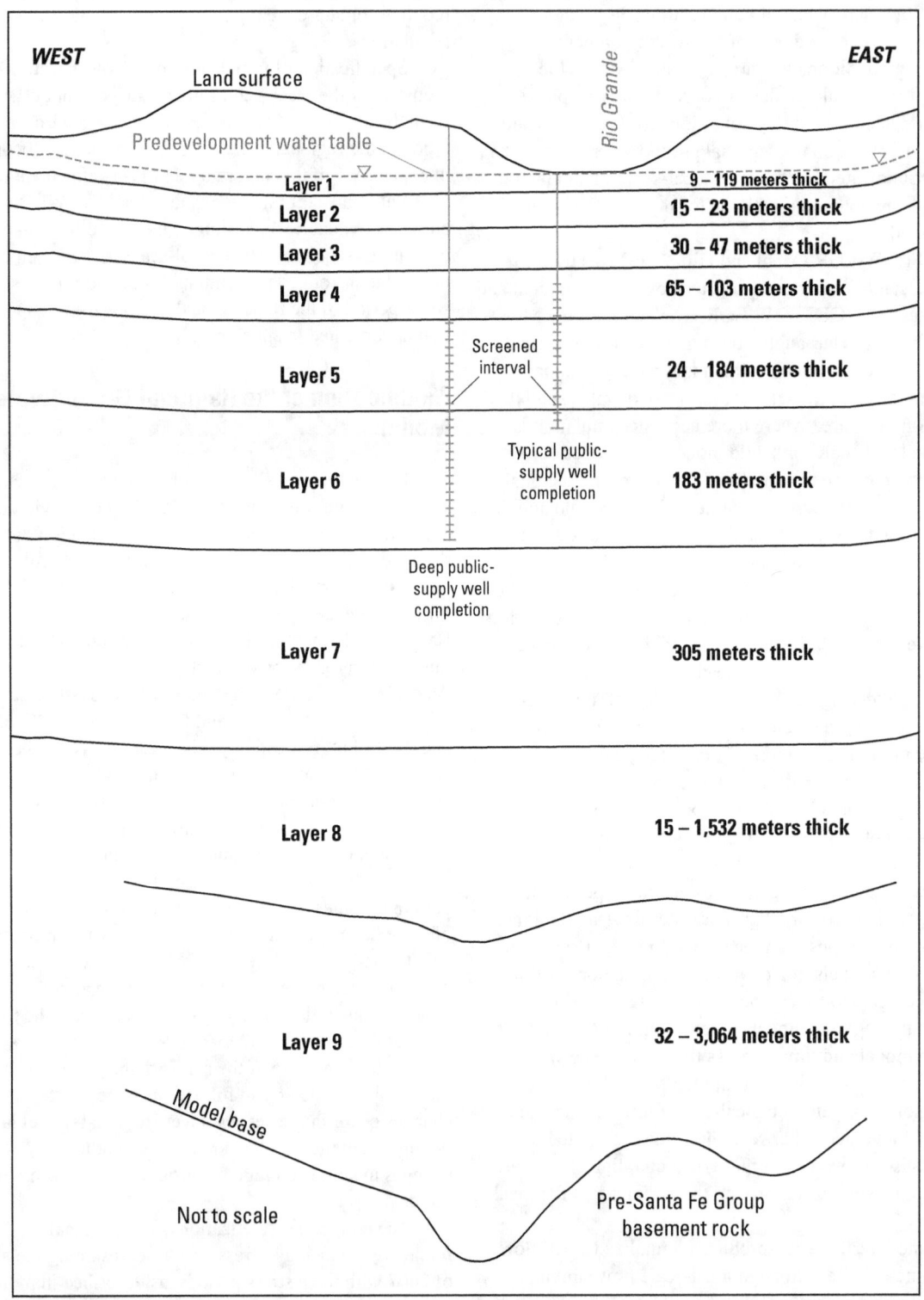

Figure 5. Layer-thickness ranges of the modified regional flow model.

represented by the hydraulic properties and stresses simulated in the corresponding local-scale model. Because hydraulic properties simulated with the local-scale model typically differ from those simulated in the corresponding portion of the regional model, the regional model is effectively modified in the refined area when coupled to the local-scale model.

Local-Scale Model Geometry and Discretization

The local-scale model fully penetrates and replaces a 12-row by 8-column portion of the regional-model domain in which groundwater flow within the 24-km^2 area of the detailed study area is simulated. With respect to the regional model, the local-scale model domain is discretized with a 5:1 horizontal refinement ratio, resulting in a 40 by 60 grid of square cells, each 100 m on a side. Each model cell therefore encompasses an area of 0.01 km^2. The 9 layers in the regional model are refined vertically into 45 child-model layers. The vertical refinement ratios of parent-model to child-model layers is variable, with no refinement of the top two and bottom three parent-model layers and substantial refinement of parent-model layers 3 and 4 (table 2). The relatively thin 3-m thickness of child-model layers 3–34 was required to simulate hydraulic-conductivity variability within the hydrogeologic-facies realizations described previously. Child-model layers 35–42 incorporate coarser vertical-refinement ratios that transition layer thicknesses to the substantial aquifer intervals represented by the unrefined layers 43–45.

The temporal discretizations of the regional- and local-scale models are identical. Conditions prior to 1900 are represented by a steady-state stress period, which is followed by transient stress periods that simulate the 109 years through December 31, 2008. Fifteen 5-year stress periods simulate the time from 1900 to 1974, and fifteen 1-year stress periods

Table 2 Child-model refinement with respect to regional model, Albuquerque Basin, New Mexico.

[Layers numbered downward from the top; m, meters]

Regional layer[1]	Vertical refinement ratio	Local layers	Local-layer thickness (m)
1	1:1	1	9.0–119
2	1:1	2	15.2–15.6
3	10:1	3–12	3.05
4	22:1	13–34	3.05
5	5:1	35–39	24.5
6	3:1	40–42	61.0
7	1:1	43	305
8	1:1	44	675–991
9	1:1	45	1351–1983

[1]Regional model layering documented in Bexfield and others (2011).

simulate the time from 1975 through 1989. Following a 74-day stress period representing the winter of 1990, seasonal stress periods representing time after March 15, 1990, generally simulate 230-day "irrigation season" periods that extend from March 16 through October 31 and 135-day (or 136-day during leap years) "winter" periods that extend from November 1 through March 15. The 5-year, 1-year, 230-day, and 135-day stress periods are divided into 8, 6, 5, and 4 time steps, respectively. The time-step lengths were specified with a geometric time-step-length multiplier of 1.5, resulting in an initial time-step length of about 17 days for the annual and seasonal stress periods.

The 136-day "winter" period that simulates conditions from November 1, 2007, through March 15, 2008, was divided into three stress periods of 31 days, 1 day, and 104 days to enable accurate simulation of non-pumping and pumping periods before and during collection of wellbore-flow data in the SSW.

Local-Scale Groundwater-Flow Model Boundary Conditions

Specified flux and head-dependent flux boundaries in the local model were used to simulate vertical recharge and groundwater withdrawals. Similar to the regional-scale model (Bexfield and others, 2011), recharge through the unsaturated zone in the area simulated by the local model was assumed to be insignificant prior to urbanization. Accordingly, recharge to and withdrawals from the local model were assumed to be zero during the steady-state stress period, which represents time prior to 1900. Anthropogenic recharge that was simulated for times after 1900 may have resulted from concentrated water sources that have the potential to wet portions of the vadose zone sufficiently to cause downward migration of water through the vadose zone, resulting in recharge at the underlying water table.

Recharge

Although the magnitude and distribution of water infiltrating to the water table is uncertain, the presence of tracers of young (post-1950s) groundwater, such as tritium and chlorofluorocarbons, suggests that such recharge does occur. Likely sources of recharge water include leakage from sewer and water-distribution pipelines, infiltration from an unlined stormwater diversion channel, infiltration beneath ponds, and infiltration of excess irrigation to urban turf and park areas. The distribution of these sources within the local-scale model area (fig. 6) was specified using zonation in the MODFLOW recharge package. The magnitude of recharge flux associated with each source was estimated by parameterization during model calibration. Although the simulated total recharge is reasonable, the relative contributions from each source, while reasonable, are not well constrained and thus uncertain.

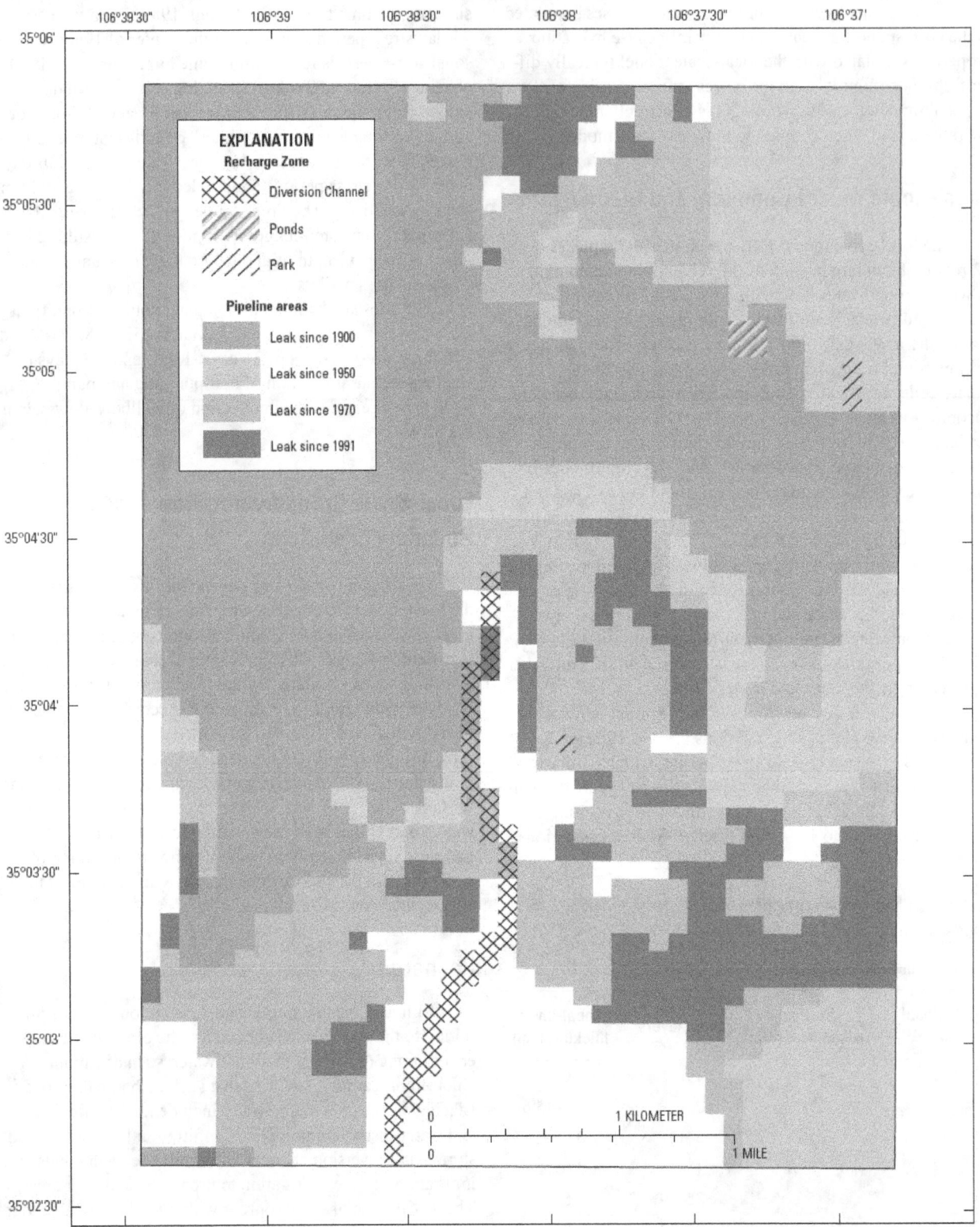

Figure 6.   Distribution of simulated recharge sources in the local-scale model, Albuquerque, New Mexico.

Leakage from sewer and water-distribution pipelines

The areas susceptible to recharge from leaky sewer and water-distribution pipelines were specified in a fashion similar to that used for the regional-scale model (Bexfield and others, 2011). Geographic information system (GIS) databases of the extent of the Albuquerque Metropolitan area in the years 1935, 1951, 1973, and 1991 (Feller and Hester, 2001) were intersected with GIS databases of the City of Albuquerque water-distribution and sewage-pipe systems to approximate the service areas that existed during four time periods: 1900–49, 1950–69, 1970–90, and 1991–2007. All areas of service during one of the four time periods were considered equally susceptible to recharge due to leakage from the water-distribution and sewage-pipe systems. Although leakage from the water-distribution and sewage-pipe systems occurs at the discrete locations where breakage occurs and changes through time, these locations are unknown. This recharge is therefore simulated as a diffuse source over a broader area than the more focused sources that occur in reality (but with unknown locations).

The magnitude of recharge in the specified possible pipe-leakage areas was parameterized as a fraction of the total reported groundwater withdrawals for each stress period, including those in the coupled regional model. This fraction, which was constant throughout the transient simulation, was estimated by model calibration to be about 3.3 percent. The recharge simulated from leaky pipes was about 600 cubic meters per day (m³/d), or 13 percent of the total simulated recharge within the local model for the year 1988. In comparison to the simulated pipe leakage, during a water-loss analysis (New Mexico Environmental Finance Center, 2006) for the Albuquerque Bernalillo County Water Utility Authority (ABCWUA), it was estimated that about 18,650 m³/d of "unaccounted-for water" may be due to "undetectable" and "detectable" distribution leakage throughout the ABCWUA service area, which was about 4.8 percent of the ABCWUA groundwater withdrawals during 2005.

Channel Leakage

The South Diversion Channel in Albuquerque was constructed in 1968 (Peterson, 1992) to provide flood protection by diverting stormwater to Tijeras Arroyo (fig. 1). Within the local model area, the unlined portion of this channel from which leakage is simulated during and after the year 1970 is depicted in figure 6. Although a gage exists where the diversion channel discharges into Tijeras Arroyo, there is not a gage on the upper reach of the channel with which to calculate flow loss along the channel. The average leakage from the channel within the local model area was estimated during model calibration to be 1,297 m³/d, or about 0.5 cubic feet per second (ft³/s). This leakage equates to an average flux on the order of 10 centimeters per day over the unlined area of the channel. For comparison, measured discharge from the channel at Tijeras Arroyo between June 8, 1988, and March 25, 2008, varied from 0 to 230 ft³/s. Although the recharge simulated

from this channel was about 29 percent of the total simulated recharge into the local model for 1988, this quantity is poorly constrained and likely smaller in reality, with a correspondingly greater portion of recharge actually coming from leaking pipes.

Infiltration from Ponds and Urban Turf Areas

The location of the duck pond west of the library on the University of New Mexico Campus, which was simulated as a source of recharge, is shown in figure 6. The vegetated areas in several parks, fields, and other urban turf areas, which are occasionally watered, are also depicted. The recharge simulated from ponds and urban turf areas was about 58 percent of the total simulated recharge into the local model for 1988.

Groundwater Withdrawals

Groundwater-withdrawal records were obtained from the New Mexico Office of the State Engineer, ABCWUA, and Bjorklund and Maxwell (1961). Because groundwater-withdrawal data prior to the 1960s were limited, earlier withdrawal rates from ABCWUA and University of New Mexico wells were extrapolated from later records (Kernodle and others, 1995). Withdrawals from other commercial wells were specified only in years for which records were available. Consequently, model-simulated withdrawals may under-represent actual withdrawals.

Reported historical groundwater withdrawals since 1900 from 33 commercial and public-supply wells within the local model area were simulated as head-dependent flux boundaries with the revised Multi-Node Well Package (MNW2) of MODFLOW (Konikow and others, 2009). Use of MNW2 was preferred over the specified-flux Well Package (WEL) in this simulation because the simulated screened intervals of the withdrawal wells spanned many finite-difference cells, among which differences in simulated hydraulic conductivity exist. Although the total reported withdrawal is specified for each well with MNW2, the layer-by-layer distribution of each specified withdrawal depends on the hydraulic conductivities and heads in each of the finite-difference cells connected to the simulated withdrawal well.

The contribution from each of the three major public-supply wells within the local study area to the total daily withdrawal from the three wells is depicted with a different color in figure 7, which illustrates well-field operation over a typical 1.5-year period from July 1, 2007, to December 31, 2008. Turbine pumps in the public-supply wells typically run at a constant rate for the time required to supply the daily withdrawal from the well. During times of peak demand, a supply well such as PSW1 may be in operation constantly for many days. Public-supply wells may also be in a non-pumping status for extended periods during times of reduced demand or if operations are limited by water-quality factors or other constraints.

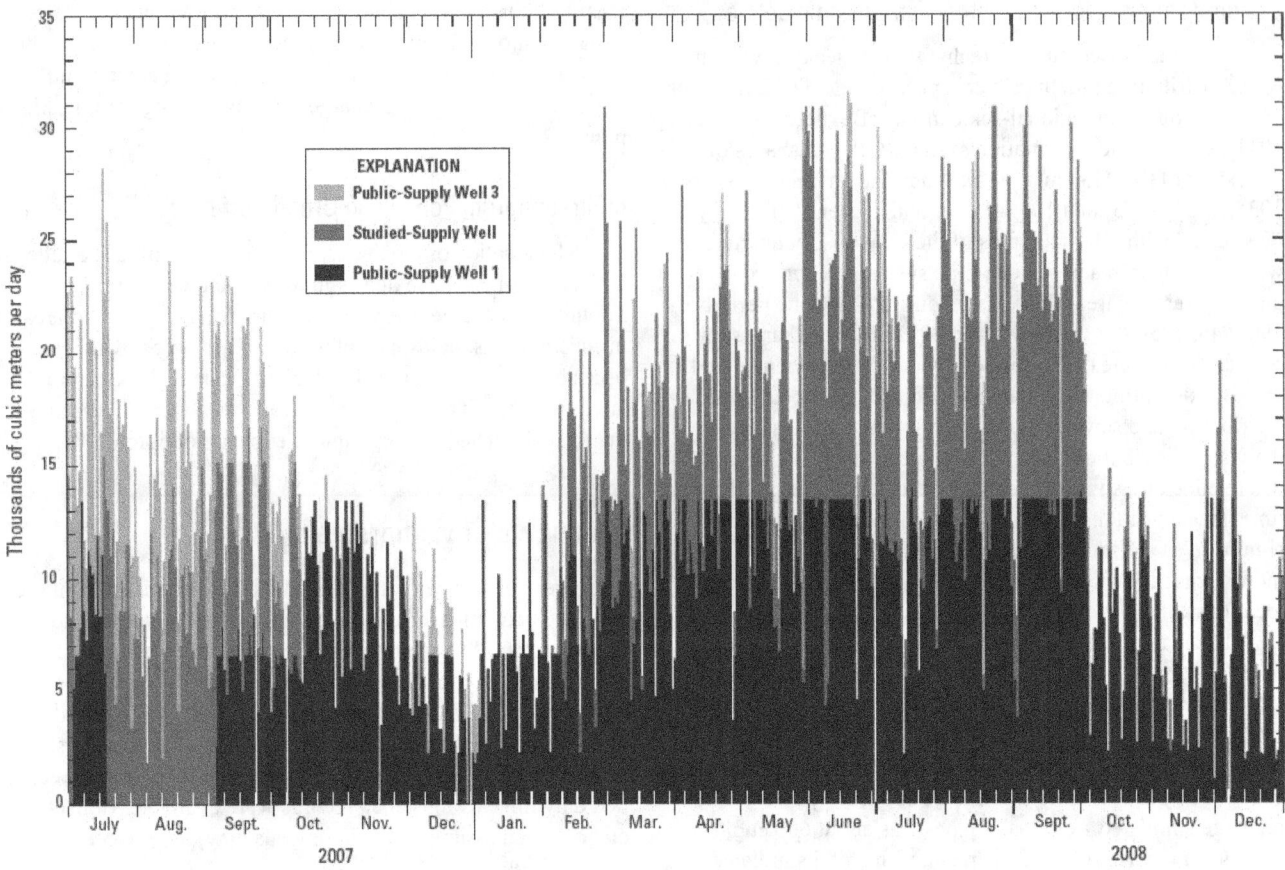

Figure 7.   Stacked column chart of daily withdrawals from public-supply wells in the local study area.

The total quantity of reported withdrawals that were simulated in the local-scale and regional (exclusive of the local-scale domain) models for each stress period is depicted in figure 8. Until about 1940, all of the reported withdrawals from the MRGB were within the Albuquerque area and were encompassed by the local-scale model domain. The effect of urban growth and consequent expansion of municipal well fields and industrial and other wells into areas surrounding the local study area is shown by the greater proportion of pumpage from the regional model (exclusive of the local-scale domain) after about 1950.

Zonation of Simulated Hydraulic Properties

The spatial distribution of vertical and horizontal hydraulic conductivities control the initial (steady-state) hydraulic head configuration, and, together with specific storage and specific yield, the response to transient stresses, such as pumpage and recharge. Model parameters representing these hydraulic properties were initially assigned values equal to those in the corresponding portion of the regional model. Within the portion of the model domain simulated with the

TPROGS hydrogeologic-facies realization, values of hydraulic conductivities for the coarse and fine hydrofacies were constrained to be higher and lower, respectively, than the intermediate facies. Estimation of the values of parameters representing hydraulic properties within the local model is discussed in a subsequent section of the report on model calibration.

Local-Scale Transport Simulation

The transport of carbon-14 (^{14}C), tritium (^{3}H), and three chlorofluorocarbon (CFC) species—trichlorofluoromethane (CFC-11), dichlorodifluoromethane (CFC-12), and trichlorotrifluoromethane (CFC-113)—was simulated with the groundwater solute-transport code MT3DMS (Zheng and Wang, 1999; Zheng, 2010). Measurements of the concentrations of each of these species in groundwater samples, in addition to measured water-level data, were used to simultaneously calibrate the local-scale groundwater flow (MODFLOW) and the transport (MT3DMS) simulations.

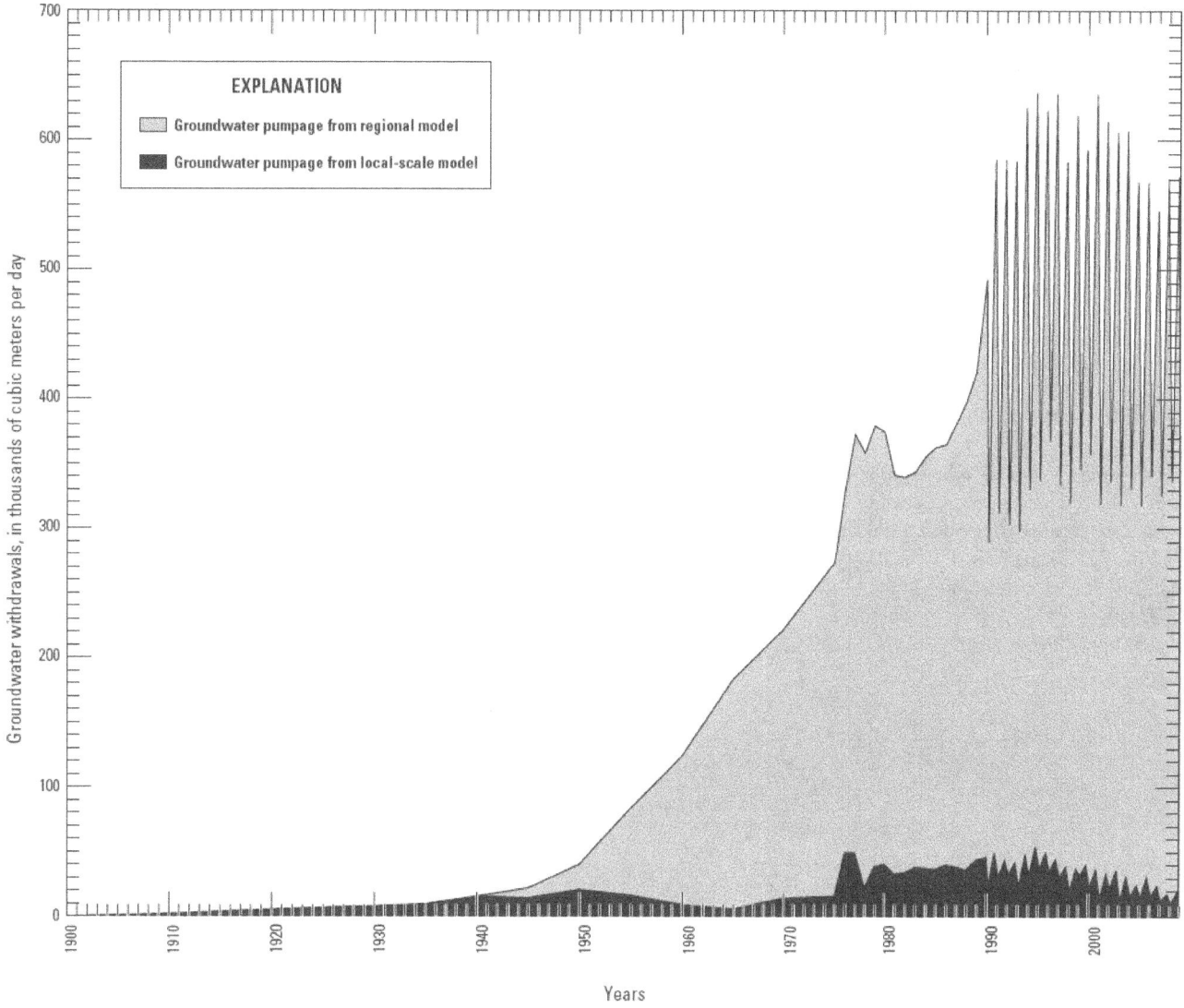

Figure 8. Reported groundwater withdrawals within the local-scale model domain and within the regional-model domain (exclusive of the local-scale model domain).

Governing Equation

MT3DMS numerically solves the transient advection-dispersion equation:

$$\frac{\partial\left(\theta C^{k}\right)}{\partial t} = \frac{\partial}{\partial x_{i}}\left(\theta D_{ij}\frac{\partial C^{k}}{\partial x_{i}}\right) - \frac{\partial}{\partial x_{i}}\left(\theta v_{i}C^{k}\right) + q_{s}C_{s}^{k} + R_{n} \quad , \quad (1)$$

where

θ	is the effective porosity,
C^{k}	is the concentration of the dissolved chemical species k,
t	is time,
x	is distance,

i and j	are	indices referencing a Cartesian coordinate system,
D	is	the hydrodynamic dispersion tensor,
v_{i}	is	the seepage velocity,
q_{s}	is	the volumetric flow rate of fluid sources and sinks,
C_{s}^{k}	is	the concentration of dissolved chemical species k in the fluid sources or sink, and
R_{n}	is	a chemical reaction term.

The advection term of equation 1 was solved with the mass conservative third-order total-variation-diminishing (TVD) scheme of MT3DMS (Zheng and Wang, 1999). Although some numerical dispersion occurs in the solution of equation 1 with TVD, the calibration of model-simulated to observed concentrations was not improved with simulation of additional hydrodynamic dispersion. Accordingly, the dispersion package of MT3DMS was not used to simulate additional dispersion.

Because the three CFC species (CFC-11, CFC-12, and CFC-113) are not known to decay significantly in the oxic groundwater conditions which exist along most transport paths from the water table to sampling wells, they were simulated as conservative tracers, with no reaction term (R_n in equation 1). For the simulation of ^{14}C and ^{3}H, the reaction term represents their respective radioactive decays.

Specified-Concentration Boundary Conditions

The concentrations of ^{14}C, ^{3}H, and each CFC species in water that infiltrates into the vadose zone near the ground surface were assumed to be equilibrated with their respective atmospheric concentrations, each of which are known to vary with time because of variable production rates (for ^{14}C, ^{3}H, and the CFCs) and radioactive decay (for ^{14}C and ^{3}H). Because the CFC transport rate through the vadose zone by gaseous diffusion was unknown, it was assumed to be less than 1 year. Accordingly, the atmospheric concentrations of CFCs (fig. 9)

were used to specify the CFC concentrations in recharge at the water table. In contrast, ^{3}H, which is a component of infiltrating water molecules, and ^{14}C, which is present as inorganic carbon dissolved within infiltrating water, likely require a longer time interval to transit the vadose zone before entering the saturated zone in recharge at the water table. During this time interval, the concentration of ^{3}H, which has a half-life of 12.32 years, may substantially decrease due to radioactive decay. Although the change in ^{14}C concentration due to radioactive decay is relatively minor during this time interval, the concentrations of ^{3}H and ^{14}C at the water table may differ significantly from their atmospheric concentrations because the atmospheric concentrations of ^{14}C (fig. 10) and ^{3}H (fig. 11) are strongly time dependent due to the effects of atmospheric nuclear testing. The effect of this delay on simulated concentrations was tested for vadose-zone transit times of zero, 10, and 20 years by comparing simulated concentrations from simulations using specified ^{14}C and ^{3}H recharge concentrations equal to the atmospheric concentration (for an "instantaneous" vadose-zone transit time) and corrected for radioactive decay during 10- and 20-year vadose-zone transit times. Although actual transit times likely vary across the study area, a uniform delay was simulated for the entire model domain. Because

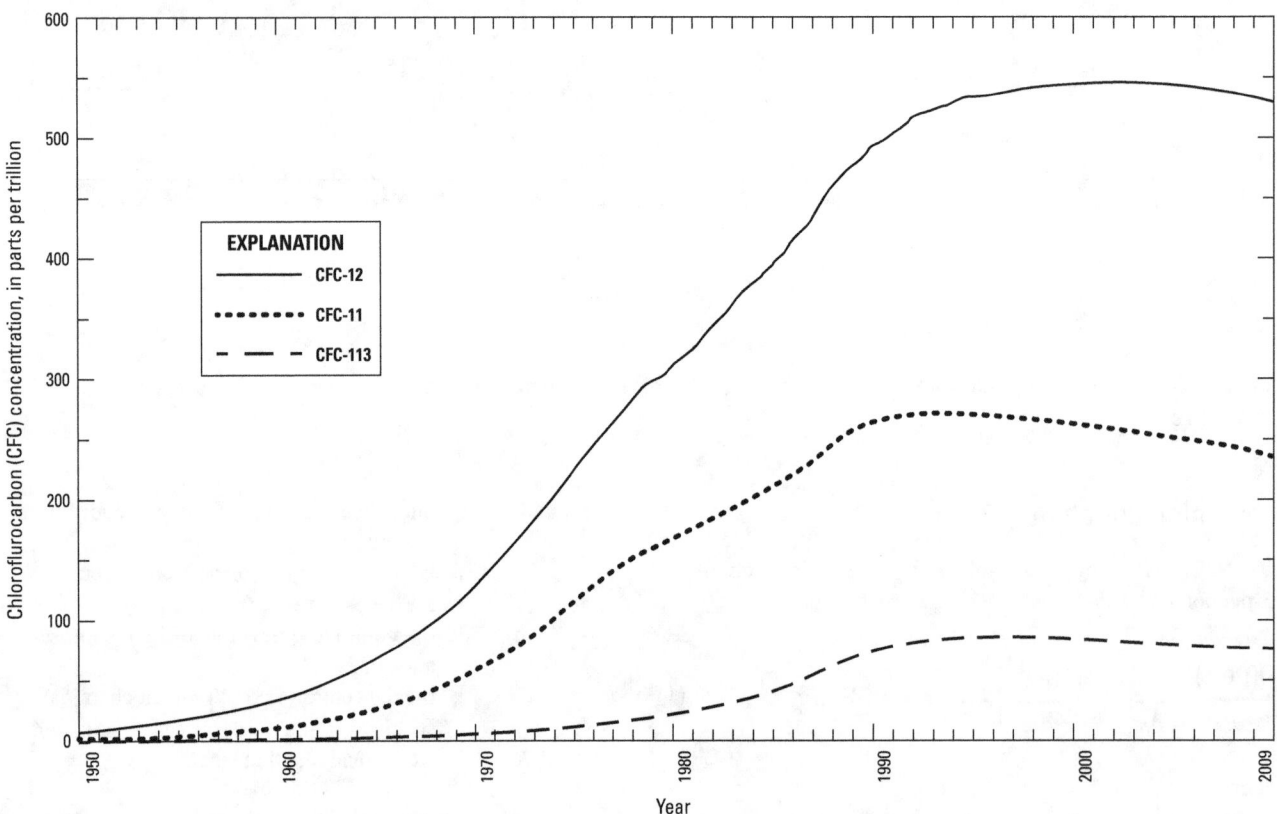

Figure 9.   Atmospheric concentrations of chlorofluorocarbons (CFC-11, CFC-12, and CFC-113) above Albuquerque, New Mexico, 1950–2010.

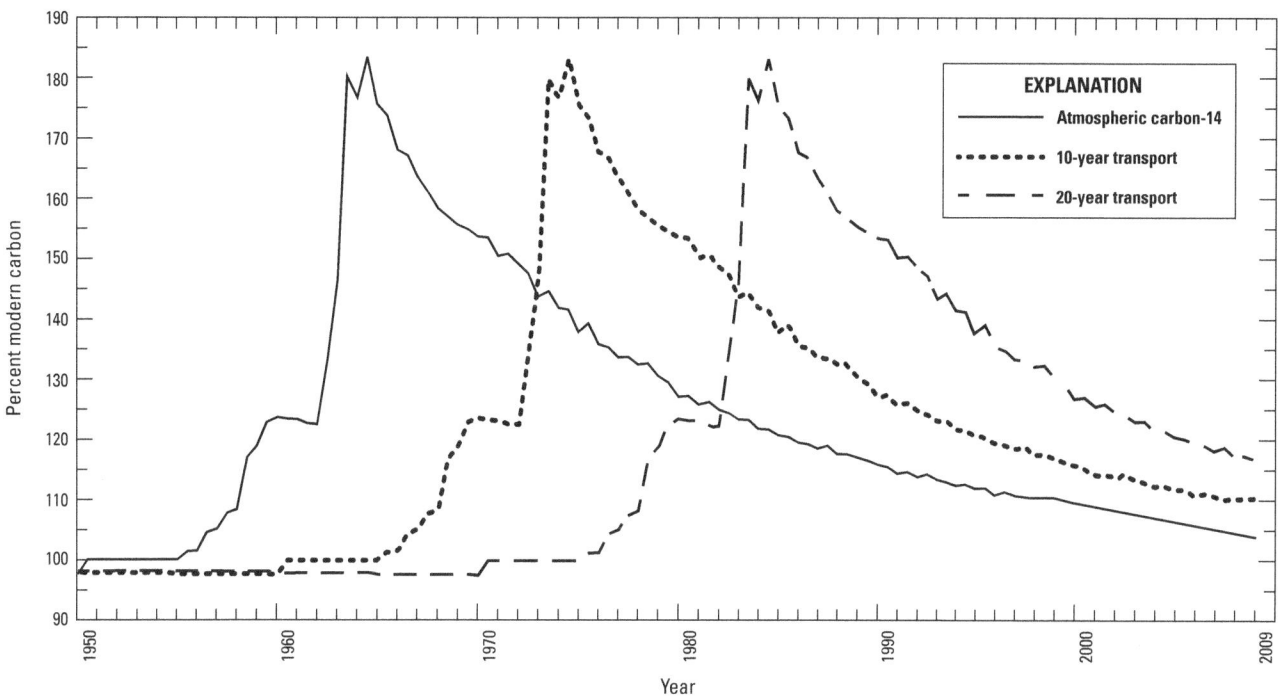

Figure 10.   Atmospheric carbon-14 concentration and specified concentrations for 10- and 20-year transport to the water table, Albuquerque, New Mexico, 1950–2009.

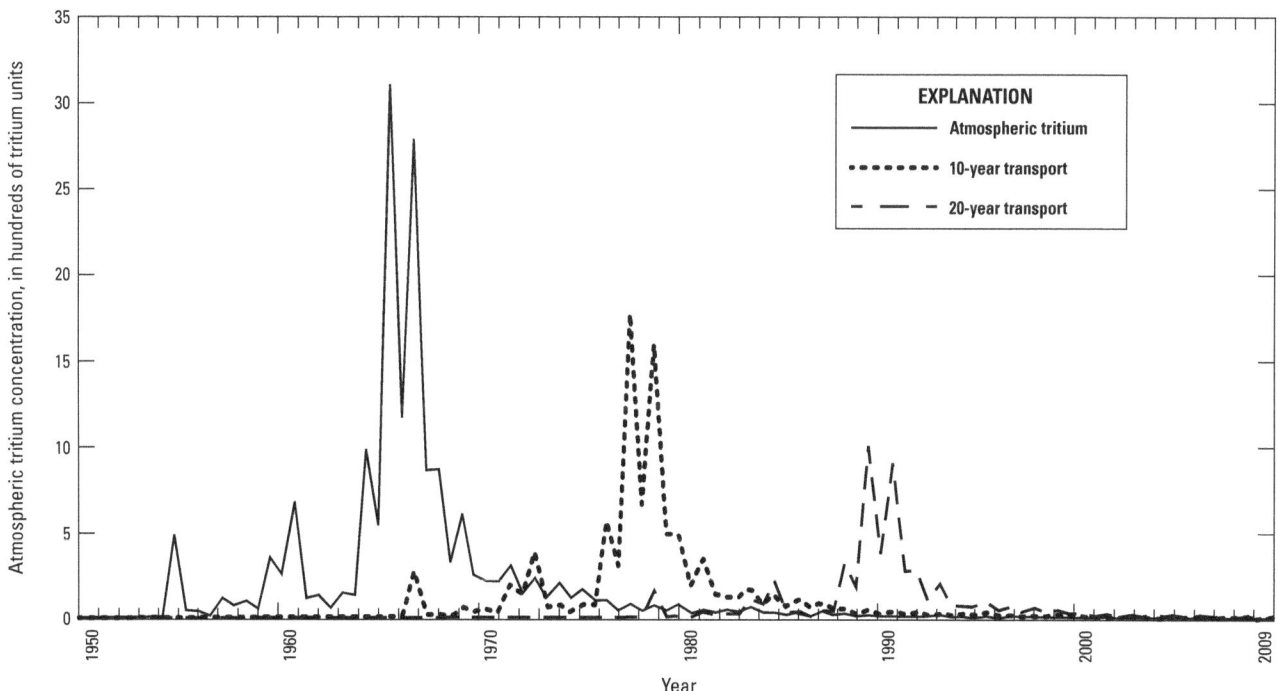

Figure 11.   Atmospheric tritium concentration above Albuquerque, New Mexico, and specified concentration for 10- and 20-year transport to the water table.

parameters representing delay time and the effective porosity in the saturated zone were correlated, the delay time could not be reliably determined through model calibration. Bexfield and others (2012) estimated that unsaturated zone travel times ranged from 8 to 20 years in the local study area. A delay time of 10 years was used to specify the ^{14}C and ^{3}H concentrations in recharge in this simulation.

Specified-Concentration Initial Conditions

Initial concentrations for the three CFC species (CFC-11, CFC-12, and CFC-113), which are anthropogenic and synthesized after 1945, were simply specified as equal to zero throughout the model domain. Prior to 1945, the "pre bomb era" background atmospheric ^{3}H concentration was about 4 parts per trillion (ppt; Michel, 1989). The subsequent radioactive decay of this small background concentration while it was transported through the vadose and saturated zones ensured concentrations were negligible throughout the aquifer system. Accordingly, the initial ^{3}H concentration was also specified as equal to zero throughout the model domain.

^{14}C concentrations implying groundwater residence times up to about 50,000 years have been measured in groundwater samples from wells in the MRGB (Sanford and others, 2004). The predevelopment distribution of ^{14}C concentration throughout the local study area is unknown but likely was spatially variable and a function of the groundwater residence time at each location. The importance of specifying a "reasonably accurate" ^{14}C initial condition throughout the local-scale model domain motivated the regional ^{14}C transport simulation documented in appendix 1. ^{14}C concentrations within the local-scale model domain were interpolated from the simulated concentrations for the steady-state stress period of the regional simulation, which represents carbon-14 concentration during predevelopment time. The simulated initial carbon-14 concentration distribution within the top regional and local-scale model layers (fig. 12), which contain the predevelopment water-table results from the simulated convergence of *relatively* young water (still generally more than 1,000 years old) that recharged from the Rio Grande (to the north and west) and at the base of the Sandia Mountains (to the east). The shallow portion of the local study area contains a zone of mixing of waters that originated from these sources, with older waters simulated near the middle and southern parts of the local-scale model domain.

The simulated initial carbon-14 concentrations from regional model layer 5 (fig. 13), which corresponds to local-scale model layers 35–39 and simulates the aquifer interval from about 125 to 246 m deep, are much lower than the simulated initial ^{14}C concentrations in the top layer of the model, reflecting longer simulated groundwater residence times at greater depths. Younger water that is simulated as recharging along the eastern mountains also enters the eastern side of the local-scale model domain at these greater depths, but younger water from the Rio Grande does not appear in layer 5.

Model Calibration

The local-scale flow and solute-transport models were simultaneously calibrated to obtain the best overall agreement between measured heads and age-tracer concentrations and their corresponding values simulated with MODFLOW and MT3DMS, respectively. This calibration was accomplished by a non-linear regression procedure that adjusted the model-parameter values to minimize the overall differences between the observed quantities and their simulated equivalents; the sum of those differences is quantified by the objective function (eq. 2). More specifically, the parameter-estimation code PEST (Doherty, 2004) adjusted the local-scale flow and transport-model parameters in table 3 in order to minimize the objective function:

$$\sum_{j=1}^{5} \sum_{i=1}^{N} \left(\omega_i C_i - \omega_i C'_i \right)^2 + \sum_{k=1}^{172} \left(\omega_k H_k - \omega_k H'_k \right)^2 , \quad (2)$$

where

H	is	the measured water level for observation k,
H'	is	the simulated-equivalent water level for observation k,
C	is	the concentration measured for observation i of chemical species j,
C'_i	is	the simulated-equivalent concentration for observation i of chemical species j,
ω_i	is	the weight applied to concentration observation i and its simulated equivalent,
ω_k	is	the weight applied to head observation k,
N	is	the number of observations for each of the five chemical species simulated with MT3DMS, and
172	is	the number of water-level observations.

Concentration Observations

Concentrations of the five chemical species (tracers of groundwater age) measured in samples from wells (fig. 14) were used for model calibration. Bexfield and others (2012) categorized the sampled wells into four categories based on well type (monitoring or public supply) and the well-screened interval depths (table 4). The monitoring-well screen lengths are all less than 21.3 m, and most are less than 12.2 m. The shallow monitoring wells are screened across the water table or have a screen midpoint within 18.3 m of the water level in the well. The 13 intermediate monitoring wells have a screen midpoint between 27.1 and 79.5 m below the water level in the well. The two deep monitoring wells have a screen midpoint 185 m or more below the water level in the well. The three public-supply wells have screened intervals at least 190 m in length that include the depth intervals characterized by the intermediate and deep monitoring wells. The 25 wells sampled for these concentration observations compose the majority of the 28 wells from which water-level observations were used in

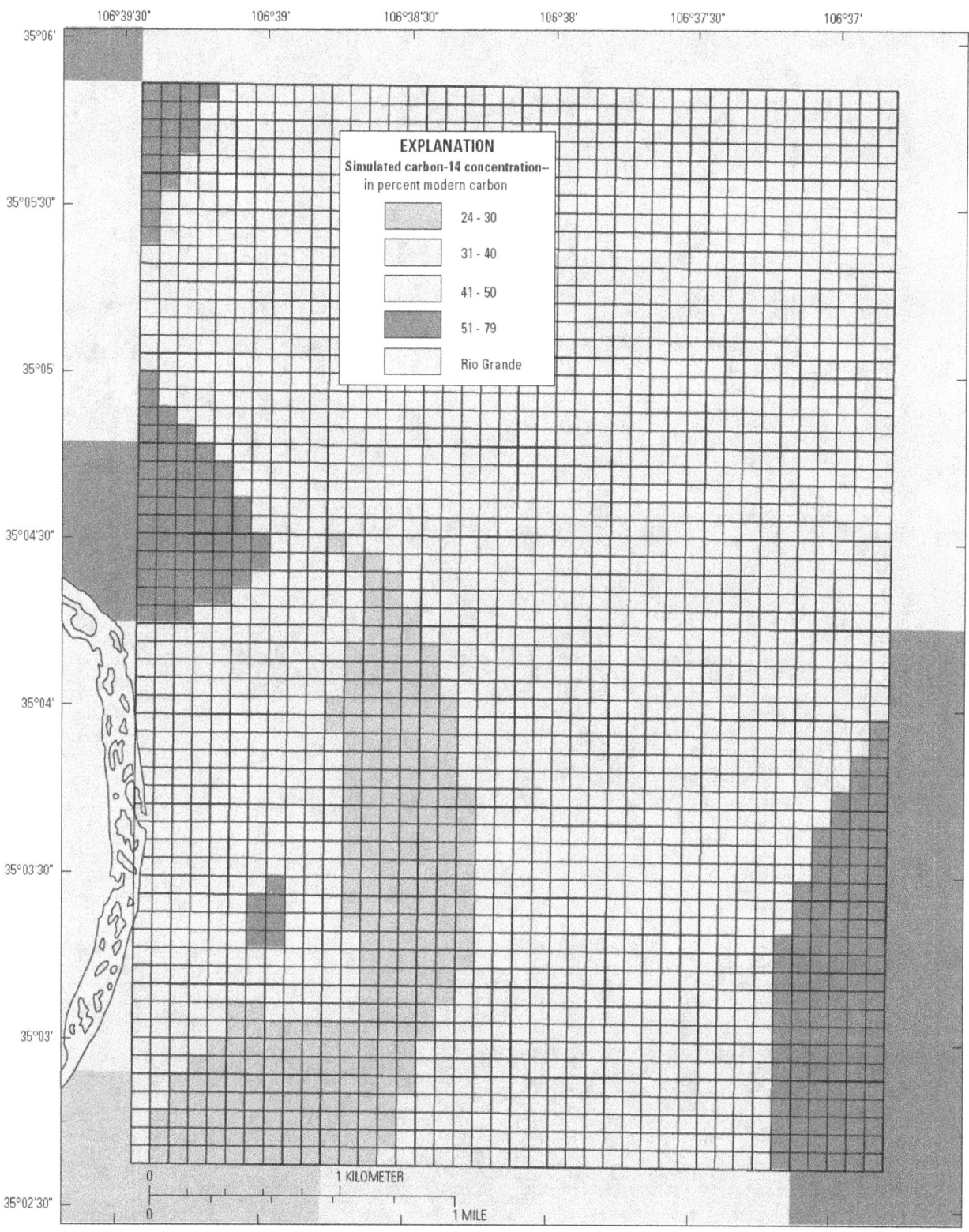

Figure 12. Regional and local-scale finite difference cells, with carbon-14 concentration interpolated from the regional transport simulation and specified as initial concentration for the top layer of the local-scale model, Albuquerque, New Mexico.

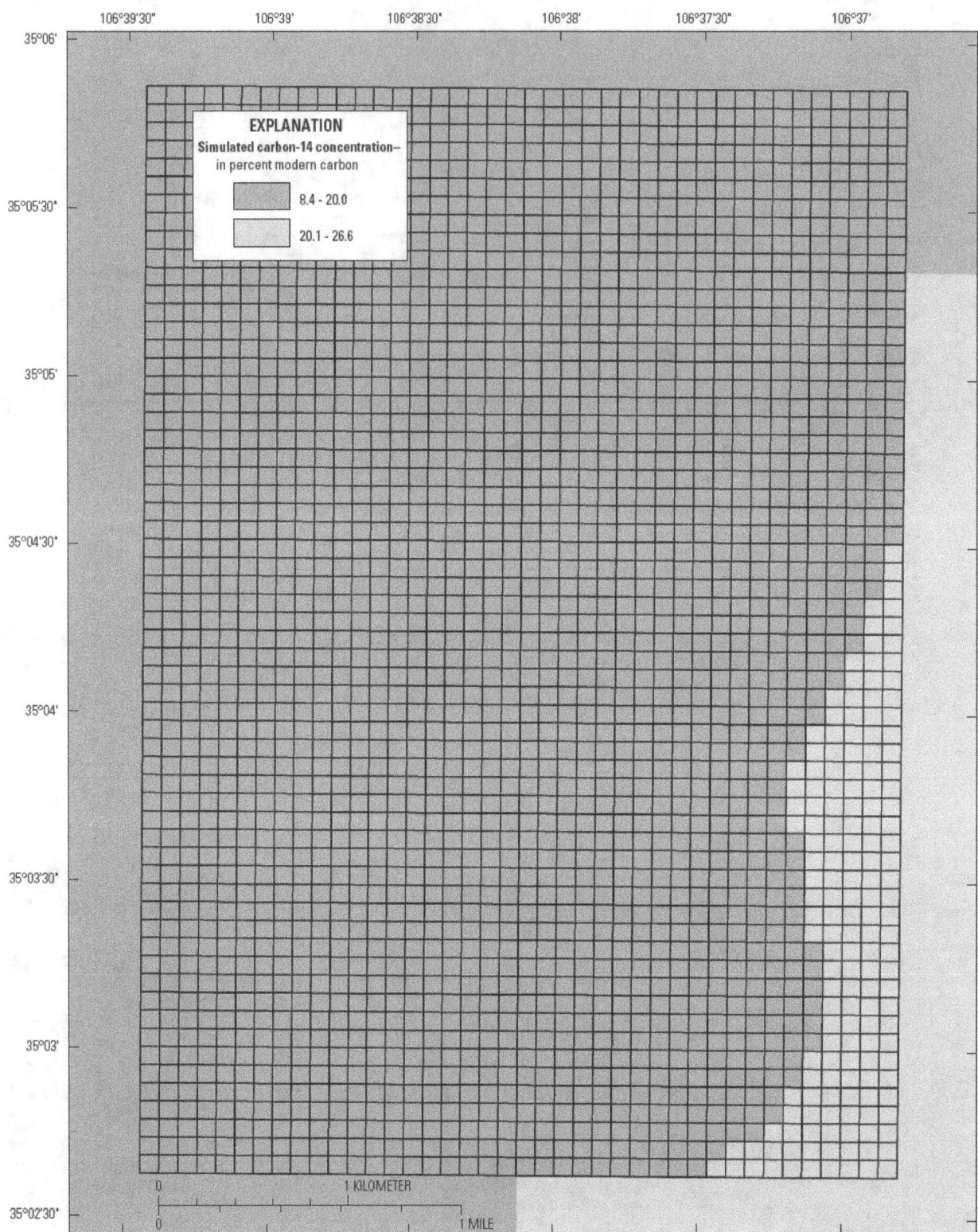

Figure 13.   Carbon-14 concentration interpolated from layer 5 of the regional transport simulation and specified as initial concentration for layers 35–39 of the local-scale model, Albuquerque, New Mexico.

Table 3. Summary of parameters representing hydraulic property values in the local scale model in Albuquerque, New Mexico.

[HK, horizontal hydraulic conductivity; VK, vertical hydraulic conductivity; HANI, horizontal anisotropy (ratio of HK along columns to HK along rows); VANI, vertical anisotropy (ratio of VK to specified HK); SS, specific storage; SY, specific yield; n, effective porosity; Fm, Formation; m/d, meters per day; m^{-1}, per meter]

Parameter name	Type	Description	Initial value	Calibrated value
K_Qr	HK	River alluvium in layers 1, 2	2.4 m/d	8.5 m/d
K_QTsa	HK	Sierra Ladrones Fm. in layers 1, 2	13.7 m/d	22.7 m/d
K_fine	HK	Fine facies in layers 3–34	0.3 m/d	0.23 m/d
K_moderate	HK	Intermediate facies in layers 3–34	7 m/d	16.9 m/d
K_coarse	HK	Coarse facies in layers 3–34	46 m/d	121 m/d
K_Tcrp	HK	Ceja Fm: Rio Puerco member in layers 35, 36, 38, 39	2.4 m/d	1.22 m/d
K_Tcrp2	HK	Ceja Fm: Rio Puerco member in layers 37, 40	2.4 m/d	3.35 m/d
Ksdm	HK	Ceja Fm: Atrisco member in layers 41–44	0.46 m/d	0.46 m/d
Ksdfm	HK	Fine Sand in layer 45	0.015 m/d	0.015 m/d
Haniyes	HANI	Layers 1, 2, 35–45	2:1	2.66 : 1
Hanino	HANI	Layers 3–34	1:1	1:1
Vani1	VANI	Layers 1, 2	1:1	39:1
Vani2	VANI	Layers 35–45	15:1	13:1
VK_coarse	VK	Coarse facies in Layers 3–34	9.1 m/d	34.3 m/d
VK_moderate	VK	Moderate facies in Layers 3–34	0.3 m/d	0.76 m/d
VK_fine	VK	Fine facies in Layers 3–34	0.03 m/d	0.09 m/d
SSall	SS	All layers	6.6×10^{-6} m^{-1}	5.25×10^{-6} m^{-1}
SYall	SY	All layers	0.2	0.12
porosity	n	All layers	0.215	0.08

model calibration. Details of the sampling and analysis of the chemical species are documented in the companion report by Bexfield and others (2012).

Observations of carbon-14 and tritium concentrations from all of the 25 sampled wells were used. CFC concentration observations denoted as "not determined" in table 4 were not used for model calibration because either (1) the measured concentration was higher than what is possible from atmospheric input, indicating contamination from other CFC sources, (2) the ratio of a CFC to other CFC species indicated that it likely was contaminated, although its concentration was not higher than possible from atmospheric input, or (3)

the measured CFC concentration was very low compared to measurements of the other CFCs in the same well, indicating degradation of that species.

The estimated errors associated with each concentration observation were used to assign the weights (ω_i) in the objective function (eq. 2) used for the model calibration. For each chemical species, various error estimates were converted to a measurement variance, the inverse of which defined the initial measurement weight. The weights within a group of measurements of each chemical species were then multiplied by a factor to correct for different measurement units. Thus, concentration observations with greater measurement or laboratory

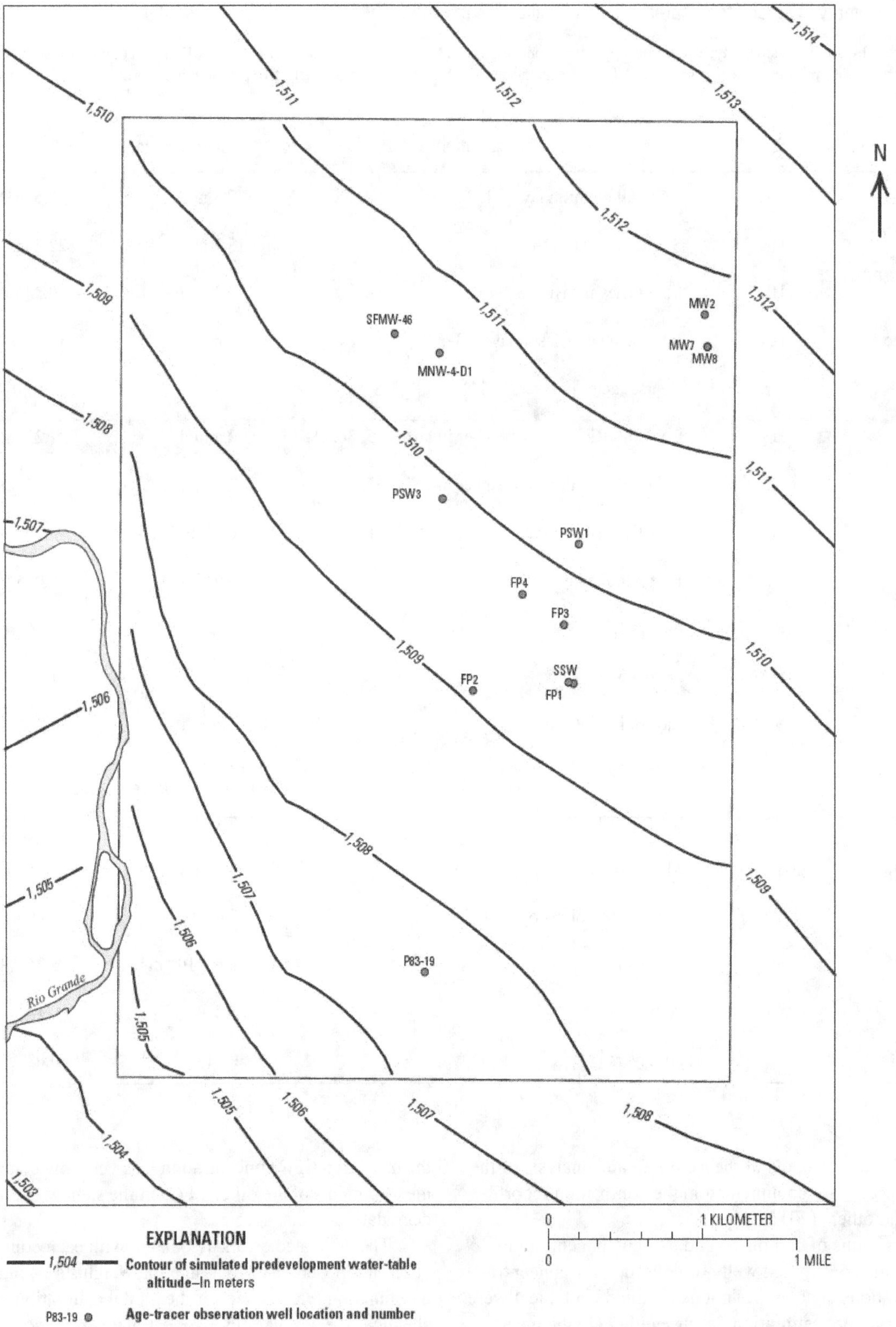

Figure 14. Boundary of the local-scale model, locations of age-tracer observation wells, and predevelopment water-table altitudes simulated with the local-scale flow model, and with the coupled regional-scale flow model, Albuquerque, New Mexico. (See table 4 for observation well information.)

Table 4. Concentration observations used to calibrate the local-scale flow and transport model in Albuquerque, New Mexico.

[14C, carbon-14; pmc, percent modern carbon; TU, tritium units; CFC, chlorofluorocarbon; pptv, parts per trillion by volume; DD, depth-dependent sample, followed by depth of collection; WH, wellhead; sqrwt, Square root of weight; nd, not determined; m, meters]

Well name (fig. 14)	Well category	Sample date (mm/dd/yyyy)	Sample time sample type	14C activity (pmc)	14C (sqrwt)	Tritium concentration (TU)	Tritium (sqrwt)	CFC-11 concentration (pptv)	CFC-11 (sqrwt)	CFC-12 concentration (pptv)	CFC-12 (sqrwt)	CFC-113 concentration (pptv)	CFC-113 (sqrwt)
FP1D	Deep	6/10/2007	1400 Regular	11.71	1.3	0.02	2.75	2.9	0.37	nd	nd	3.3	0.49
FP1MD	Deep	6/8/2007	1400 Regular	15.04	1.18	0.00	2.75	nd	nd	nd	nd	3.0	0.29
FP1MS	Intermediate	6/6/2007	1500 Regular	51.22	0.75	0.52	2.75	70.1	0.02	nd	nd	20.5	0.07
FP1MS	Intermediate	11/10/2008	1130 Regular	43.86	0.75	1.19	2.75	nd	nd	113.2	0.02	3.8	0.12
FP2D	Intermediate	6/17/2007	1500 Regular	72.85	0.64	6.03	1.25	nd	nd	67.0	0.03	0.0	0.36
FP2M	Intermediate	6/16/2007	1400 Regular	86.65	0.57	5.51	1.38	37.0	0.04	nd	nd	7.8	0.21
FP3D	Intermediate	6/14/2007	1400 Regular	68.06	0.67	7.79	0.95	nd	nd	nd	nd	19.4	0.08
FP3D	Intermediate	11/13/2008	1400 Regular	73.21	0.67	6.63	1.15	nd	nd	159.8	0.01	4.5	0.36
FP3M	Intermediate	6/13/2007	1500 Regular	59.61	0.66	0.04	2.75	23.0	0.07	8.2	0.06	5.9	0.24
FP4D	Intermediate	11/24/2008	1215 Regular	67.73	0.79	7.12	1.08	46.1	0.04	nd	nd	6.7	0.19
FP4M	Intermediate	11/25/2008	930 Regular	67.48	0.69	4.07	1.08	203.7	0.01	nd	nd	46.8	0.04
MW7	Intermediate	12/18/2007	1500 Regular	52.85	0.71	0.02	2.75	1.1	1.46	42.2	0.03	0.1	0.35
MW8	Intermediate	12/18/2007	1045 Regular	57.73	0.68	0.17	2.75	nd	nd	nd	nd	71.7	0.02
MNW4-D1	Intermediate	2/20/2008	1630 Regular	56.29	0.79	1.40	2.75	nd	nd	nd	nd	6.3	0.08
SFMW-46	Intermediate	1/15/2008	1145 Regular	71.60	1.42	4.10	1.78	nd	nd	nd	nd	8.4	0.12
P83-19LR	Intermediate	1/30/2008	1530 Regular	30.65	1.54	2.24	2.75	nd	nd	nd	nd	nd	nd
P83-19M	Intermediate	1/31/2008	1300 Regular	66.35	0.57	4.90	1.58	nd	nd	nd	nd	nd	nd
FP1S	Shallow	6/5/2007	1400 Regular	67.03	0.65	2.26	2.75	nd	nd	nd	nd	36.8	0.05
FP2S	Shallow	6/15/2007	1500 Regular	75.06	0.98	5.73	1.3	116.1	0.01	nd	nd	37.0	0.05
FP3S	Shallow	6/12/2007	1300 Regular	50.22	0.83	0.25	2.75	nd	nd	nd	nd	nd	nd
FP3S	Shallow	11/7/2008	1500 Regular	48.59	0.75	0.13	2.75	nd	nd	nd	nd	nd	nd
FP4S	Shallow	11/25/2008	1445 Regular	55.34	0.66	1.46	2.75	nd	nd	nd	nd	nd	nd
MW2	Shallow	12/19/2007	1100 Regular	68.26	0.94	1.65	2.75	nd	nd	nd	nd	66.3	0.03
MW9	Shallow	1/21/2008	1445 Regular	51.69	1.26	0.84	2.75	nd	nd	nd	nd	35.5	0.05
P83-19U	Shallow	1/29/2008	1315 Regular	68.30	1.06	5.08	1.48	nd	nd	nd	nd	nd	nd
PSW1	Public supply	11/12/2008	1500 Regular	43.71	0.83	1.35	2.75	nd	nd	nd	nd	38.5	0.04
PSW3	Public supply	11/13/2008	930 Regular	10.12	1.4	0.12	2.75	4.3	0.03	10.6	0.04	1.7	0.33
SSW	Public supply	6/7/2007	1400 Regular	43.77	0.79	2.35	2.75	nd	nd	192.2	0.01	5.4	0.14
SSW	Public supply	11/10/2008	930 Regular	31.20	0.88	0.96	2.75	nd	nd	33.2	0.05	0.6	0.33
SSW	Public supply	12/4/2007	1200 DD, 115 m	21.99	1.05	0.54	2.75	nd	nd	15.8	0.1	0.1	1.14
SSW	Public supply	12/4/2007	1400 DD, 133 m	21.51	1.03	0.64	2.75	nd	nd	18.0	0.08	0.5	1.05
SSW	Public supply	12/4/2007	1600 DD, 151 m	25.23	0.94	0.77	2.75	nd	nd	86.3	0.02	10.5	0.08
SSW	Public supply	12/4/2007	1700 DD, 176 m	11.65	1.42	0.14	2.75	nd	nd	28.3	0.06	3.6	0.16
SSW	Public supply	12/5/2007	1300 DD, 241 m	9.69	1.54	0.06	2.75	nd	nd	0.0	0.17	0.3	5.57
SSW	Public supply	12/5/2007	1500 DD, WH	24.49	1.06	0.70	2.75	nd	nd	7.6	0.22	0.3	0.31

analytical error had smaller weights in the parameter-estimation regression. Tritium weights were calculated from the measurement uncertainties reported by Bexfield and others (2012, table 10), which are largely attributable to laboratory analytical error. Carbon-14 weights were calculated from variances associated with analytical error and the geochemical modeling by Bexfield and others (2012), which introduced the majority of the uncertainty in the "observed" carbon-14 values. The estimated variance for each modeled carbon-14 observation was based on two assumptions: (1) a normal distribution and (2) a 90-percent confidence that the true value lies between the measured value and the "best" modeled value. CFC weights were calculated from variances associated with analytical error and the uncertainty in the recharge temperature and altitude for calculation of the "observed" CFC atmospheric mixing ratios reported by Bexfield and others (2012).

Water-Level Observations

Measurements of water levels from the public-supply, multiple-well nests (FP1, FP2, FP3, and FP4), and other wells within the local study area (fig. 15) were compiled for model calibration. Wells lacking screened-interval information or with only one water-level measurement were not used. Measurements from the final set of 28 observation wells were sampled to construct a calibration dataset with a maximum-length period of record for each well and no more than one water-level observation from each well per model stress period.

The uncertainties associated with measurement of well-head altitudes and the depth to water in the observation well were converted to variances and summed to obtain an error variance associated with each water-level measurement. The inverse of these composite variances were used for the weights (ω_k) in the objective function (eq. 2) used for the model calibration. Several water-level observations from wells with long screened intervals were assigned a "penalty" to lower the measurement weight because of uncertainty in the effect of water levels at different aquifer depths on the composite measured water level.

For observation wells with screens that span multiple model layers, the Head Observation Package (HOB) of MODFLOW requires specification of the relative contribution of the simulated head in each model layer to the simulated equivalent water level for the observation well. Because this information was unknown, the relative contributions of each model-layer finite-difference cell connected to an observation well screen were assumed to be equal during model calibration. In order to test the effect of the initial simplifying assumption, the cell-to-well conductance (CWC) for each cell connected to an observation well with a relatively long screened interval (PSW3) was computed by using the calibrated model parameters, and a revised model-layer apportionment was incorporated into a new HOB input package. For the tested observation well, the maximum difference between the composite water levels simulated by using these two approaches was 17 centimeters (cm).

The model-calibration regression was subsequently rerun to obtain the final model parameter set.

Calibrated Model Parameter Values

The calibrated parameter values representing the horizontal hydraulic conductivities of the fine and intermediate TPROGS facies are lower and higher, respectively, than the horizontal hydraulic conductivities reported in the Aquifer-Pumping Test section of this report. Because the pumping test withdrew water from all facies connected to the pumping-well screen and the measured water-level changes used for analysis of the test might depend upon the hydraulic conductivities of all these facies, hydraulic conductivities estimated from the test might be expected to have values that represent an aggregate of hydraulic conductivities of the constituent facies.

Sensitivity analysis is inherent in model calibration with a program such as PEST, which computes the sensitivity of each model parameter being estimated during each parameter-estimation iteration. At the calibrated model parameter values, simulated-equivalent values to the observation dataset were most sensitive to parameters representing effective porosity and recharge.

Simulated Water Levels

The local-grid-refinement method ensures continuity of simulated heads across the interface between the local- and regional-scale models. Contours of predevelopment water-table altitudes simulated for steady-state conditions within the local-scale model align with contours simulated with the coupled regional-scale model (fig. 14). Prior to the onset of groundwater pumping, groundwater flowed from the northeast toward the southwest through the local-model area, and some groundwater discharged to the Rio Grande, located immediately west of the local study area. The direction of groundwater flow simulated in deeper model layers is similar to that at the water table for predevelopment time.

Numerous public-supply and commercial wells are east of the local study area (Bexfield and others, 2012), and increased groundwater withdrawals in these areas since about 1950 have changed the direction of groundwater flow in the local study area from the predevelopment direction. Water levels simulated for times after the early 1950s indicate that the direction of groundwater flow above an altitude of about 1,200 m is generally from the west toward the east. Figure 15 depicts water-table altitudes simulated by child-model layer 37 and parent-model layer 5, which represent depths within the withdrawal-depth interval of Albuquerque public-supply wells. The finite-difference-cell centers, or nodes, in this layer of the local-scale model represent altitudes between 1,320 m and 1,329 m (near the southwest and northeast corners of the model grid, respectively). The simulated pressure head in the aquifer interval represented by this model layer varies from 157 to 177 m.

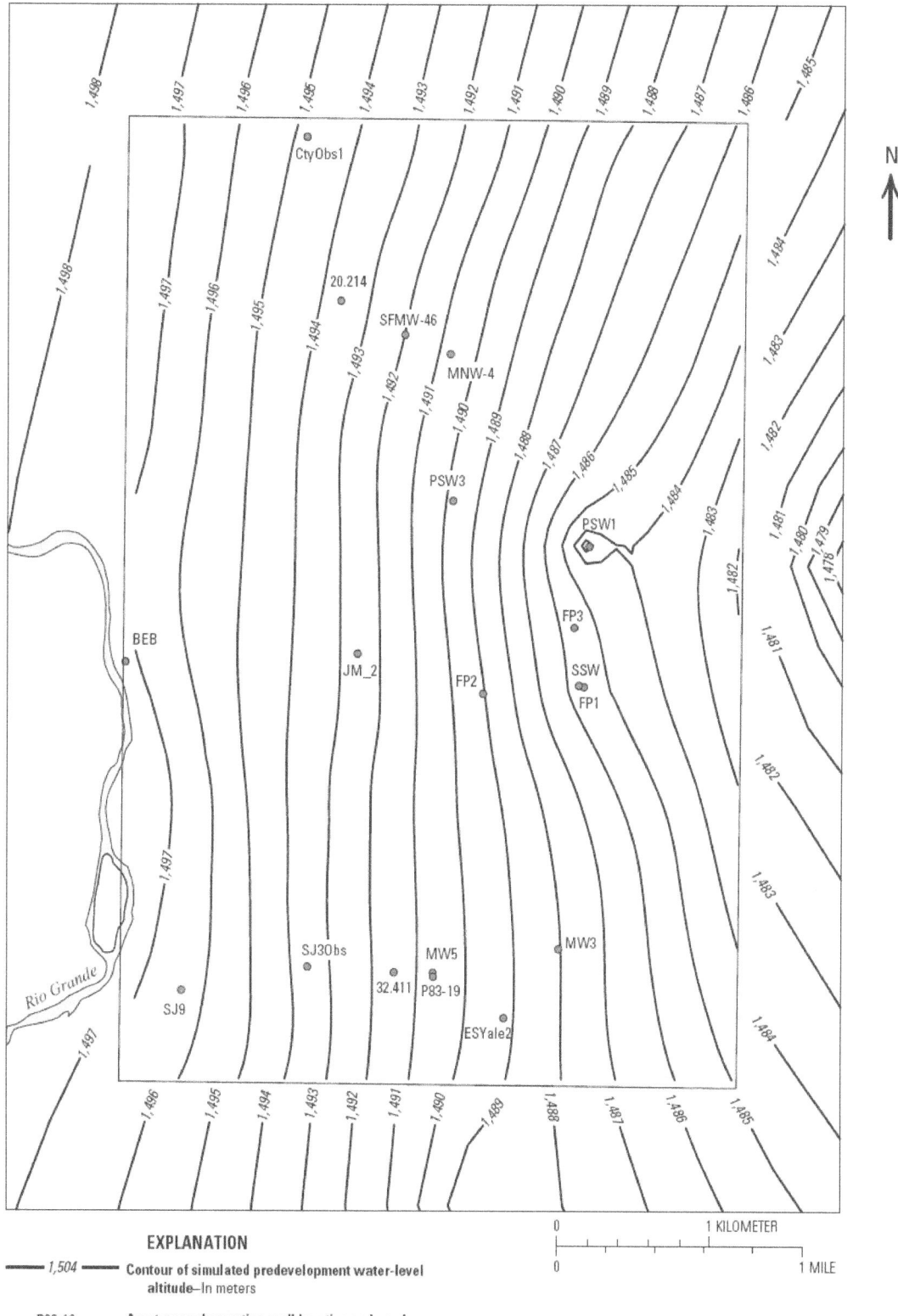

Figure 15. Boundary of the local-scale model, locations of water-level observation and public-supply wells, water-table altitudes simulated for December 31, 2009, with layer 37 of the local-scale flow model, and with layer 5 of the coupled regional-scale flow model, Albuquerque, New Mexico.

Hydrographs of observed water levels and their simulated equivalents depict the quality of the calibrated model as well as the history of water-level decline at several depths for locations within the local-model area. Observation well CityObs1, located near the north boundary of the study area, has a screen open to the aquifer in the altitude interval from 1,467 to 1,470 m, or about 43 m below land surface. Historical water levels measured in CityObs1 document water-table lowering at this location since about 1960 (fig. 16). The simulated-equivalent hydrograph for these observations, which depicts the average of water levels simulated in child model layers 8 and

9 at the model location corresponding to City Obs1, depicts trends and magnitudes of changes similar to the observations from this well.

Observation well ESYale2, near the south boundary of the study area, has a screen open to the aquifer from 76 to about 91 m below land surface. Measured water levels during the relatively short period of record in the 1980s record a lowering of the water table in this area and are simulated by the model to within 1 m.

Observation well SJ3Obs, located north of the Albuquerque International Airport in the southern part of the study area,

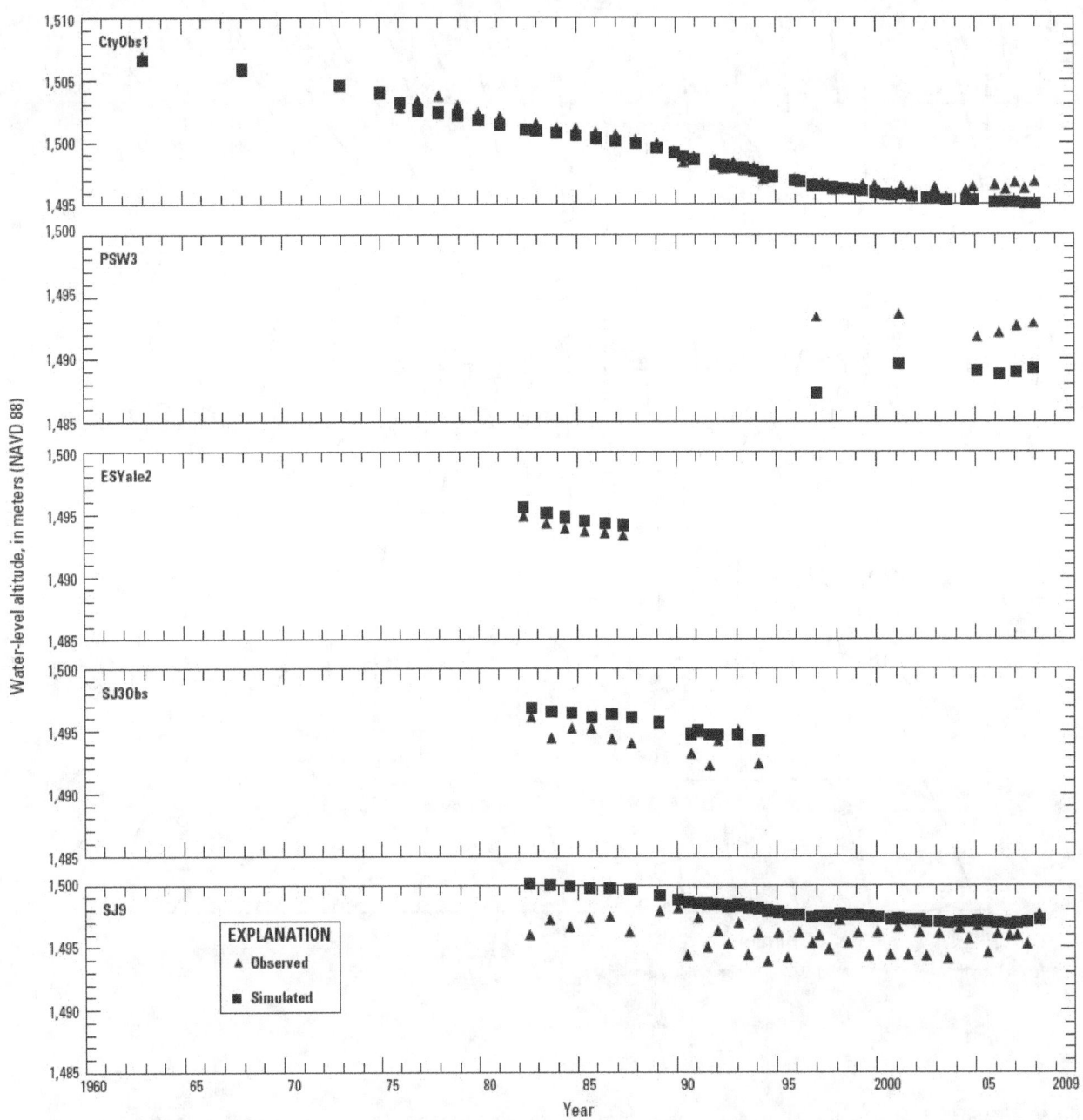

Figure 16.   Observed and simulated water levels for public-supply and observation wells in the local model, Albuquerque, New Mexico.

is screened over an interval from about 110 to 153 m below land surface. The simulated water-levels depicted in the hydrograph for well SJ3Obs are the average of the water levels simulated in the eight vertically adjacent child-model finite-difference cells corresponding to the observation well screened interval. Although the simulated hydrograph fits the observed water-level decline over the 13-year period of record, some of the observed shorter period (possibly seasonal) water-level variations are not simulated.

Observation well SJ9, in the southwest part of the study area, is screened over a relatively long interval from about 55 to 233 m below land surface. The simulated water-levels depicted in the hydrograph for this well are the average of the water levels simulated in 27 vertically adjacent child-model finite-difference cells corresponding to the observation well screened interval. The simulated hydrograph is similar to that for observation well SJ3Obs in that it adequately represents the long-term water-level decline but does not simulate the observed shorter period water-level variations, which appear to be seasonal. It is possible that water levels in observation well SJ9 respond to seasonal variations in nearby groundwater pumpage that are not accurately specified in the model or to variations in the stage of the nearby Rio Grande that may not be accurately simulated in the regional model.

Well PSW3 is a public-supply well from which withdrawals are simulated in the local-scale model. As such, there are several additional considerations in the interpretation of water levels associated with this well. Although the observed water levels are 5 m higher than their simulated equivalents, these water levels were measured after the pump in the well had been turned off for several days to weeks, allowing water levels in the well to recover somewhat toward the ambient head conditions in the surrounding aquifer. In contrast, the model-simulated water levels are calculated for the time of the observation during a model stress period in which pumpage from the well is also simulated by the model. Because this water-level recovery time cannot be simulated with the seasonal stress periods employed during the observation times, the simulated water levels are somewhat lower than their observed counterparts. The solid line in figure 17 depicts the water level simulated within the wellbore of PSW3 by MNW2 for a 34-year time period represented by 54 stress periods. This calculation of wellbore head is distinct from the "simulated equivalent" heads (depicted as square points in figure 17) that are calculated by the HOB by use of spatial and temporal interpolation between nodal heads for the location and time of the head observation, and the calculation is a more accurate representation of the head in the well while it is pumped. During stress periods in which a relatively small quantity of water is pumped, such as those for late 2008 to 2009, the MNW2-simulated wellbore head is close to that simulated by the HOB.

The fit of all simulated to observed water levels used for model calibration, including those from 23 wells not shown in the hydrographs (fig. 16), is depicted in figure 18. The reasonably random distribution of the scatterplot around the 1:1 line suggests that there is no appreciable bias in the model.

Simulated Intra-Borehole Flow Log

Depth-dependent data on borehole flow in the SSW collected on December 1 and 2, 2007, (Bexfield and others, 2012) provided an additional calibration constraint on the vertical distribution of horizontal hydraulic conductivity in the local-scale flow model. Measurements of borehole flow at discrete depths were made under "ambient" conditions, when the SSW had not been pumped for the previous 47 days. Subsequent borehole-flow measurements made several minutes after activation of the submersible pump in the well and again after several hours are plotted separately as the measured fraction of the total pumped flow as a function of depth (fig. 19). These plots depict the cumulative flow distribution during the early transient hydraulic response of the aquifer to the pumping and during a quasi-steady flow regime that may exist after pumping has continued for several hours. The greatest difference between the "early observation" and "late observation" plots occurs at shallow depths because the logging tool was trolled downward and then upward; thus, the greatest difference in time between observations at specific depths occurs at shallower depths. Whereas the plot of the "early observation" log indicates that flow is entirely into the well (fraction of the total flow decreases with depth) above a depth of 145 m shortly after pumping commences, the plot of the "late observation" log indicates that some return flow out of the well (fraction of the total flow increases and decreases with depth) occurs within this depth interval after the flow field has had sufficient time to stabilize.

Model stress periods were designed to simulate the periods of non-pumping and the intra-borehole flow test from the SSW. Head-dependent flows simulated for the time of the test into each node representing the screened interval of the SSW were used to generate the "simulated" flow log for the SSW (fig. 19), which more closely resembles the "late observation" plot. The form of the simulated cumulative-flow plot was most sensitive to the parameters K_Tcrp and K_Tcrp2 (table 3), which represent the horizontal hydraulic conductivity of the Rio Puerco member of the Ceja Formation within local-model layers 35–40 that are connected to the simulated SSW below a depth of 177 m. Because simulated SSW nodal flows were not readily incorporable into the model-calibration regression, the K_Tcrp and K_Tcrp2 parameter values were manually calibrated and subsequently constrained during the final model calibration with PEST.

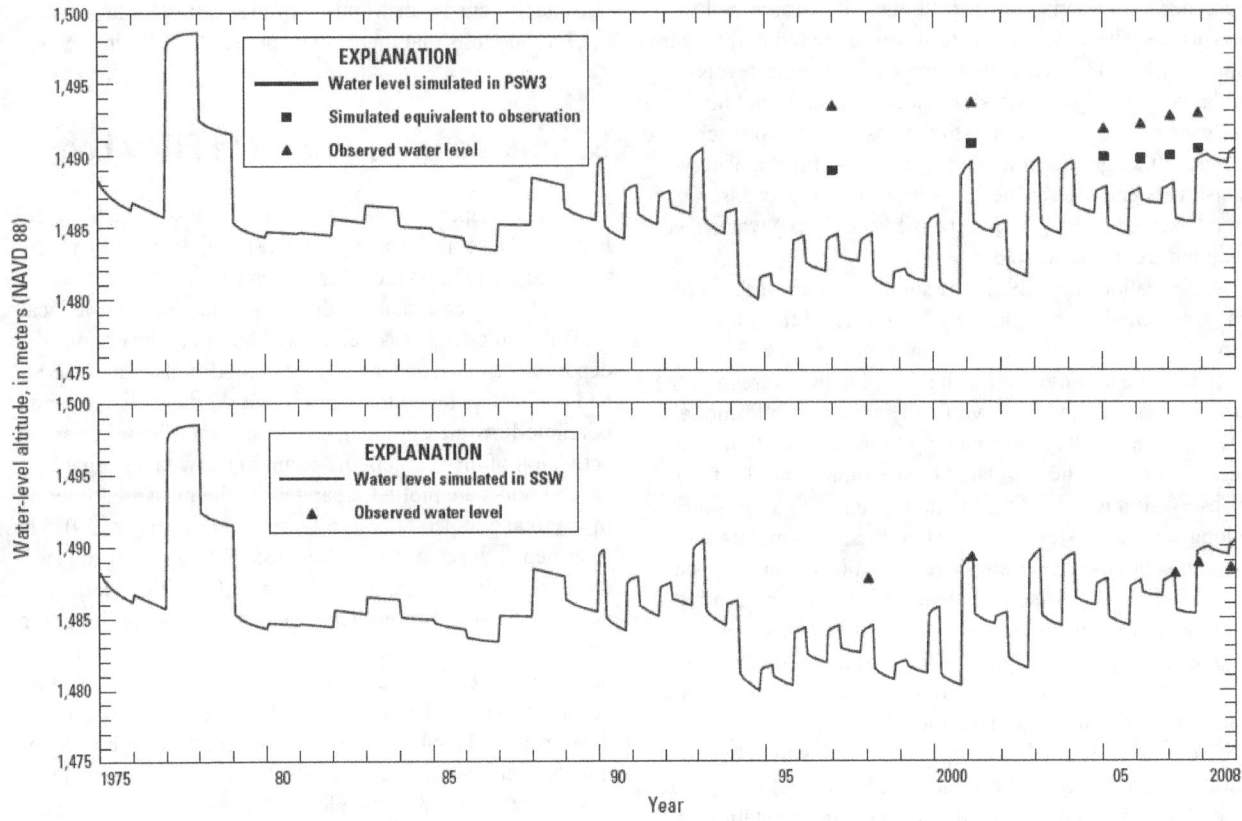

Figure 17. Water levels observed in the studied supply well (SSW) and PSW3, and simulated with the local model, Albuquerque, New Mexico.

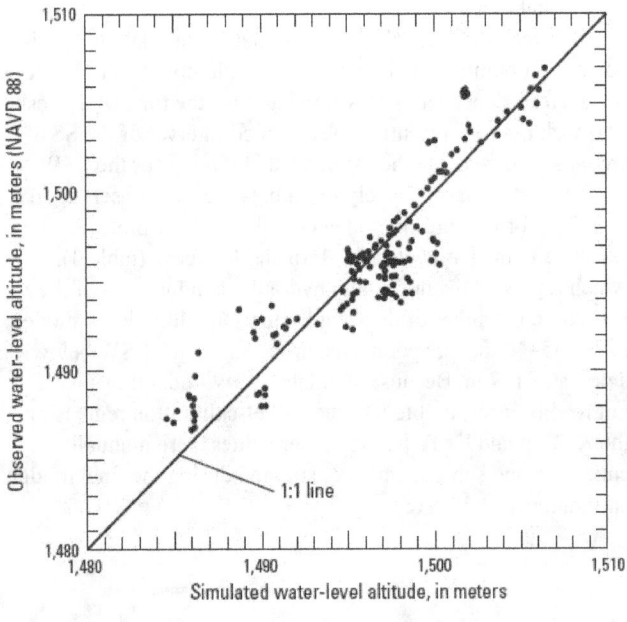

Figure 18. Relation between water levels observed and simulated with the local model, Albuquerque, New Mexico.

Simulated Concentrations of Carbon-14, Tritium, and Chlorofluorocarbons

Simulated-equivalents to observed concentrations for each of the five chemical species were computed with the Transport Observation Package (TOB) of MT3DMS. The TOB is analogous to the HOB of MODFLOW; concentrations simulated at finite-difference-cell centers are spatially interpolated for the location of the observation well and temporally interpolated between transient transport-time steps for the time of the observation. Observation wells with screens that span multiple model layers require specification of the relative contribution of the simulated concentration in each screen-penetrating model layer to the simulated equivalent concentration for the observation well. Because the relative contributions from various depth intervals to the measured concentrations were unknown, the simulated concentration contributions of each transport model-layer finite-difference cell to the simulated observation concentration were assumed to be equal.

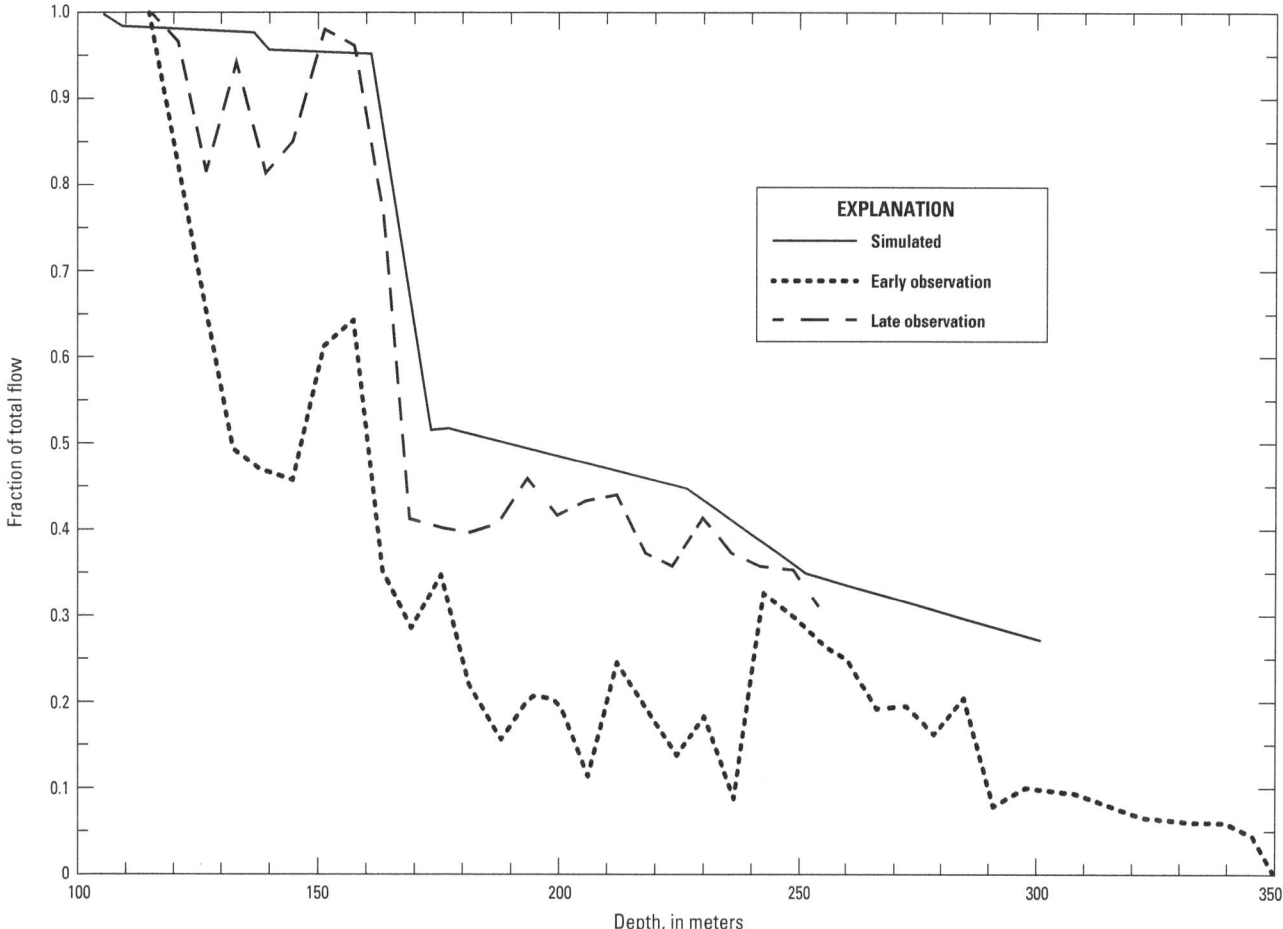

Figure 19. Simulated and observed flow logs for the studied supply well (SSW), Albuquerque, New Mexico.

Carbon-14

The graph of the relation between simulated carbon-14 concentrations and observed carbon-14 values for each well-depth category (fig. 20) illustrates that the general trend of decreasing observed carbon-14 concentrations at greater aquifer depths is well simulated. In comparison to a similar plot of observed carbon-14 concentrations and carbon-14 concentrations simulated with a regional transport simulation (fig. 1-2), simulation of carbon-14 concentrations in shallow and intermediate depth observation wells, in particular, is improved in the local-scale simulation. This improvement likely results from incorporation of the heterogeneous hydraulic conductivity distribution and additional sources of recharge to the water table in the local-scale model.

Although the carbon-14 residual pattern (fig. 20) appears random, individual depth classes of wells appear to show some bias in their residual distributions, possibly indicating a systematic model error. For example, the observed carbon-14

is typically greater than that simulated for the intermediate-depth observation wells. This bias may result from error in the carbon-14 initial condition at intermediate depths, or the existence of faster transport pathways to the intermediate-depth observation wells than are simulated by the model. (The Model Uncertainty and Limitations section discusses such pathways, which include transport through the wellbores and the gravel pack in the annular volumes surrounding the wellbores.) In contrast, the observed carbon-14 is typically less than that simulated for the public-supply wells, which may result, in part, from error in the specification of the relative contributions of model layers (assumed to be equal) to the simulated concentration for the well.

The transport code MT3DMS can simulate flux-weighted concentrations within wells with long screened intervals, such as the SSW, that were simulated as multi-node wells with the corresponding MODFLOW simulation. These concentrations simulated for multi-node wells with MNW2-simulated wellbore head differ from the TOB simulated-equivalent

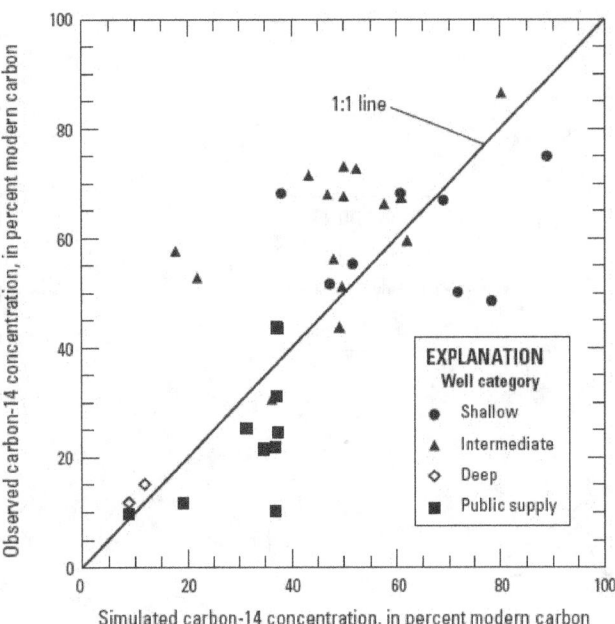

Figure 20. Relation between observed carbon-14 concentrations and concentrations simulated with the model of the local study area in Albuquerque, New Mexico.

concentrations for reasons analogous to the differences between MNW2-simulated water levels and those calculated by the HOB package of MODFLOW that were mentioned in the Simulated Water Levels section of this report. The simulated carbon-14 concentration within the SSW as a function of time is depicted with two observed carbon-14 concentrations in figure 21. Differences between seasonal groundwater withdrawal rates were simulated for periods after 1990, resulting in noticeable simulated-concentration fluctuations after that time. Carbon-14 concentrations were higher during the summer, when more water is withdrawn from the SSW. The peak simulated carbon-14 concentration appears in the SSW during 1985, which is about 21 years after the 1964 peak of atmospheric carbon-14 concentration and about 11 years after the peak concentration of the transient carbon-14 boundary condition specified for a 10-year transport-delay through the vadose zone (fig. 10).

Tritium

The fit of simulated ^{3}H concentrations to their observed values for each well-depth category is depicted in figure 22. Although there is a reasonable correspondence of simulated to observed concentrations for the deep and some of the shallow- and intermediate-depth observation wells, other tritium observations are relatively poorly simulated. In particular, some intermediate and most public-supply wells with observed ^{3}H concentrations greater than 0.4 tritium units (TU) have simulated concentrations near zero, suggesting that some

preferential pathways for young-water transport to deeper-well screened intervals may not be represented in the model.

The simulated ^{3}H concentration within the SSW is plotted with time and with three observed ^{3}H concentration measurements in figure 23. The peak in the simulated ^{3}H concentration appears in the SSW during 1981, which is about 18 years after the 1963 peak atmospheric ^{3}H concentration and 8 years after the peak concentration of the ^{3}H transient boundary condition specified for a 10-year transport delay through the vadose zone (fig. 11). The apparent 3-year difference in arrival times of the peak concentrations for ^{3}H and carbon-14 is an artifact of the flow and transport model temporal discretization; 5-year stress periods were employed prior to 1975, and specified boundary concentrations were averaged over the length of the stress period. Thus, the simulated peak concentration for ^{3}H was specified beginning in 1970, and the simulated time for arrival of peak ^{3}H concentration in the SSW is about 11 years. Because the actual time required for transport through the vadose zone is unknown and likely spatially and temporally variable, little additional uncertainty was introduced by this temporal discretization artifact.

Chlorofluorocarbons

A plot of the relation between simulated and observed CFC-113 concentrations (fig. 24) illustrates that although the model cannot precisely simulate most observed CFC concentrations, there is reasonable correspondence for many shallow and intermediate observation wells. Some observation and public-supply wells with observed CFC-113 concentrations greater than 3 pptv have simulated-equivalent concentrations near zero. As discussed previously for a similar mismatch of simulated to observed ^{3}H concentrations, the discrepancy may be due to preferential pathways for young-water transport to deeper-well screened intervals that are not represented in the model. For the same reason, the plot showing simulated and observed CFC-12 concentrations (fig. 25) also illustrates that the simulated-equivalent concentrations to many CFC-12 observations in public-supply wells are near zero. Several CFC-12 measurements from intermediate-depth observation wells are reasonably well simulated, however. Although fewer CFC-11 concentrations were deemed valid for analysis, the fit of simulated and observed CFC-11 concentrations (fig. 26) is similar to that for CFC-113 (fig. 24). A geochemical/age-dating analysis (Bexfield and others, 2012) found that although CFCs were useful in indicating the depths to which young water had been transported, the potential for CFC contamination limited the use of the CFCs in determining individual interpreted ages.

The maximum concentrations simulated in the SSW for all three CFC species appear during 1994 to 1995. Unlike the atmospheric concentrations of carbon-14 and ^{3}H, which exhibit distinct bomb-induced temporal variability (figs. 10 and 11), the specified boundary concentrations of the three CFC species vary more gradually and attain near-maximum

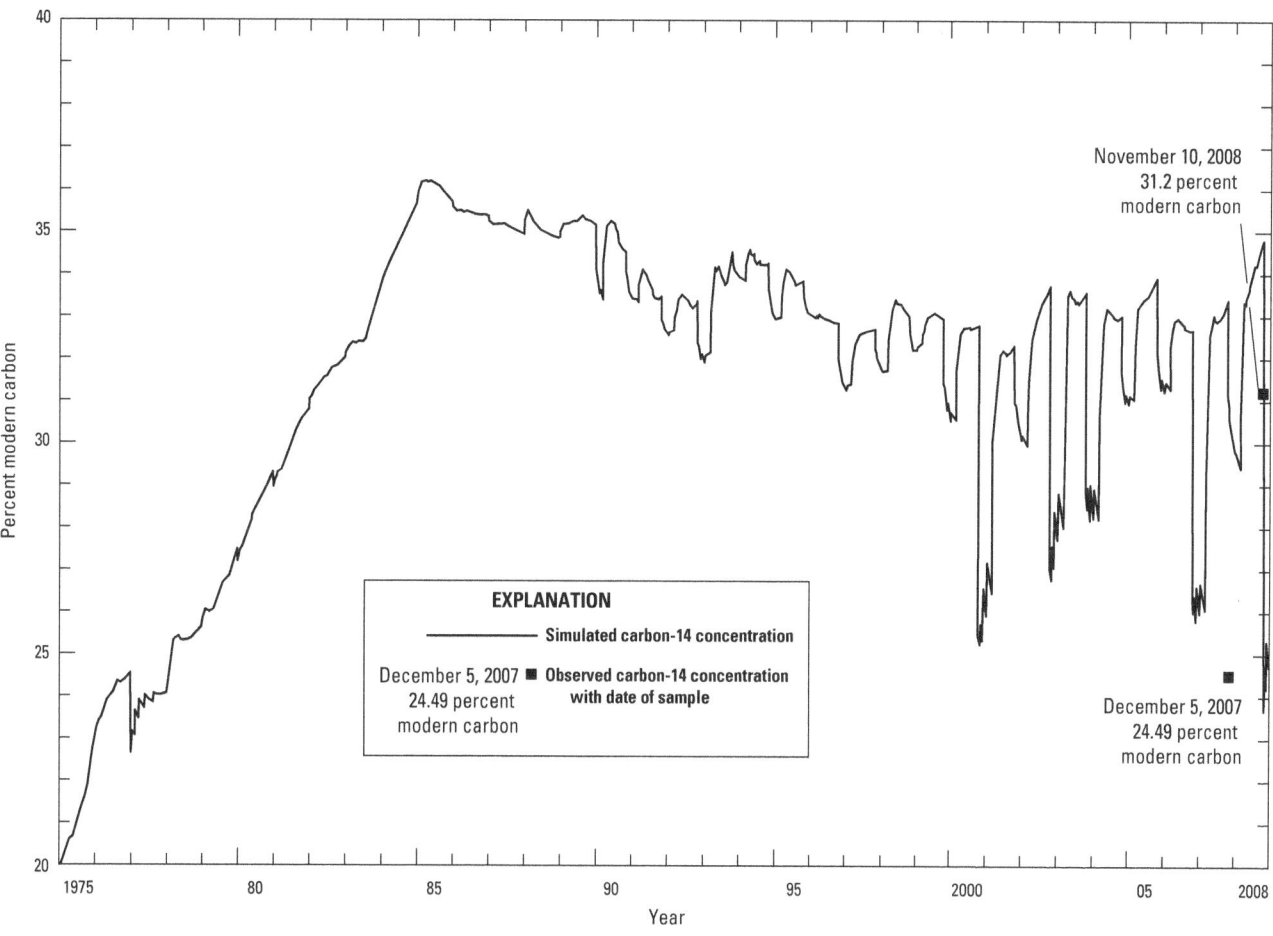

Figure 21. Carbon-14 concentrations observed and simulated in the studied supply well in Albuquerque, New Mexico.

values by about 1991 (fig. 9). Plots of the CFC species through time (fig. 27) have similar forms because of their identical source locations and transport pathways but differ in magnitude because of their respective source concentrations (fig. 9). The faster apparent transport times of simulated peak CFC concentrations during the 1990s, as compared to those for ^{3}H and carbon-14 during the 1970s, may be due to the larger groundwater withdrawal rates (fig. 8) and resulting larger hydraulic gradients in the vicinity of the SSW during the 1990s.

Simulated Binary Age Mixtures in and Surrounding the SSW

The relative proportions of young and old waters in supply wells are of interest because of the different types of contamination that may be present in water that originated from different sources or that has resided in the aquifer for different lengths of time. In the case of the SSW, the young water fraction may contain VOCs of anthropogenic origin, whereas the older fraction is more likely to be associated with the natural contaminant arsenic. Although public-supply wells with deeper screened intervals may withdraw groundwater with a smaller fraction of young water, and therefore be less susceptible to anthropogenic contaminants (such as VOCs), these wells may contain higher concentrations of natural contaminants if they are associated with older water. In the local study area, arsenic concentrations generally increase with depth (Bexfield and others, 2012), implying association with older, deeper groundwater. The fractions of "young," post-1950 groundwater, which is transported from the water table downward into the SSW screened interval, and of distinctly older groundwater that is transported upward from beneath the SSW screened interval may vary seasonally as pumping rates and consequent vertical hydraulic gradients change. Because groundwater of intermediate age surrounds the SSW, the time variation in the fractions of "young" and "old" groundwater with respect to this "intermediate" mixture were simulated, and are discussed separately.

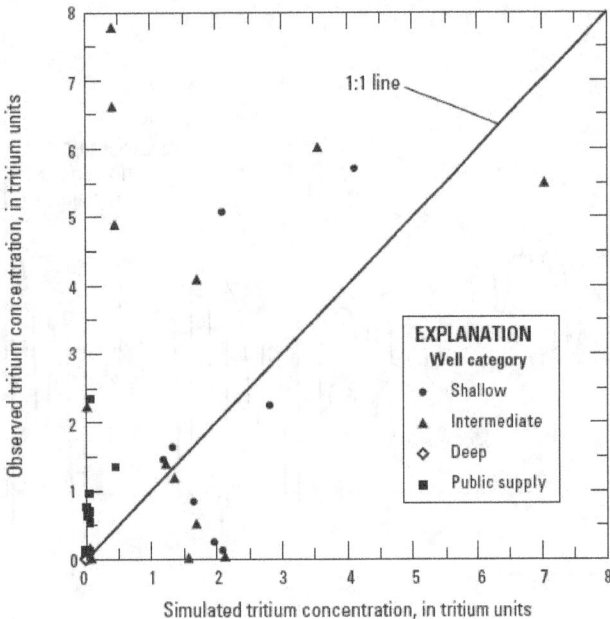

Figure 22. Relation between observed tritium concentrations and concentrations simulated with the model of the local study area in Albuquerque, New Mexico.

Simulated Fraction of Young Groundwater in and Surrounding the SSW

The time-dependent fraction of "young water" throughout the local-scale model was simulated with the transport code MT3DMS by considering "young water" as a conservative tracer that recharges the groundwater system after 1950. For this simulation, an initial zero-concentration condition was specified to represent water that resided in the saturated zone prior to 1950. The percentage of "young" water in water-supply wells that were simulated as multi-node wells, such as the SSW, can be depicted in a similar fashion as done for the five chemical tracers (figs. 21, 23, and 27). The percentage of simulated "post-1950" water in the SSW over the 38-year interval ending in 2008 (fig. 28) attained maximum values of about 12 percent between about 1985 and 1995, which is when groundwater withdrawals in the local study area were larger than in subsequent years (fig. 8).

As groundwater withdrawals from the SSW decreased somewhat from the late 1990s onward, greater differences in the simulated percentages of "young" water between seasons result from greater differences in seasonal withdrawals. Simulated percentages of young water typically vary from a low of about 4 percent during the lower-pumping wintertime periods to about 8 percent during the higher-pumping summertime periods. This seasonal difference results from greater simulated downward flow from sediments above the SSW screen during the summer pumping season. During the winter, water-level observations from the FP1 nest of piezometers,

located slightly east of the SSW (fig. 2) indicate either a smaller downward or an upward hydraulic gradient, both of which imply less flow of younger water to the SSW during the winter season. The simulated fractions of younger water generally agree favorably with those from the independent mixing analysis (Bexfield and others, 2012) that did not directly incorporate pumping rates and ranged between 3 percent and 6 percent in the winter to 11 percent in the summer.

The similarity between the simulated "post-1950" percentages (fig. 28) and the carbon-14 concentrations in the SSW (fig. 21) results from the similarity of their source boundary condition and transport pathways. A noticeable difference after about 1995 is that the relative carbon-14 concentration does not decrease as much as the relative decrease in the percentage of "post-1950" water after groundwater-withdrawal rates declined. This is due to the contribution of carbon-14 from the deeper aquifer intervals where a finite initial concentration was specified.

The simulated percentage of young, or "post-1950," water in the SSW borehole and a sampling of the model cells penetrated by the SSW is depicted over a shorter, 2.8-year interval in figure 29. The percentage of young water generally decreases with the increasing aquifer depth simulated in higher numbered model layers. The amplitude of the time-variation of the simulated "post-1950" percentage primarily depends on the hydraulic properties controlling flow (hydraulic conductivity and storage), transport (effective porosity), and, to a lesser extent, the simulated depth represented by the cell. At the location of the SSW, model cells in layers 5, 6, 7, and 30 are simulated by the highest hydraulic-conductivity, coarse facies (table 3) that is most effective in transmitting the younger fraction from the water table, resulting in larger variations of simulated "post 1950" water between stress periods with differing pumping rates. The model cells in layers 12 and 22 are simulated with the "intermediate" hydraulic-conductivity facies (table 3), and the simulated variation of the fraction of young water in those cells is smaller. Although layers 21 and 34 are simulated with the lowest hydraulic-conductivity facies (table 3), layer 21 nevertheless exhibits subdued variations in the simulated fraction of younger water. There is no substantial variation in the young-water fraction in layer 34 or deeper model layers because little of the younger water is simulated to advect to the deeper aquifer intervals. Figure 29 also illustrates the temporal correspondence between the variations in percentage of young water in the borehole of the SSW (figs. 28 and 29) and the variations within the model cells contributing water to the simulated SSW.

A borehole-flow log of the SSW (Bexfield and others, 2012) indicates that flow through the borehole occurs under ambient (non-pumping) conditions. Interpreted variations in the age distribution of waters sampled from the SSW may be due, in part, to preferential flow through the borehole or gravel-filled annular space of the SSW during periods of low or no pumping. Because substantial withdrawals occurred from the SSW during each of the seasonal stress periods simulated by the transient simulation documented in this report, no

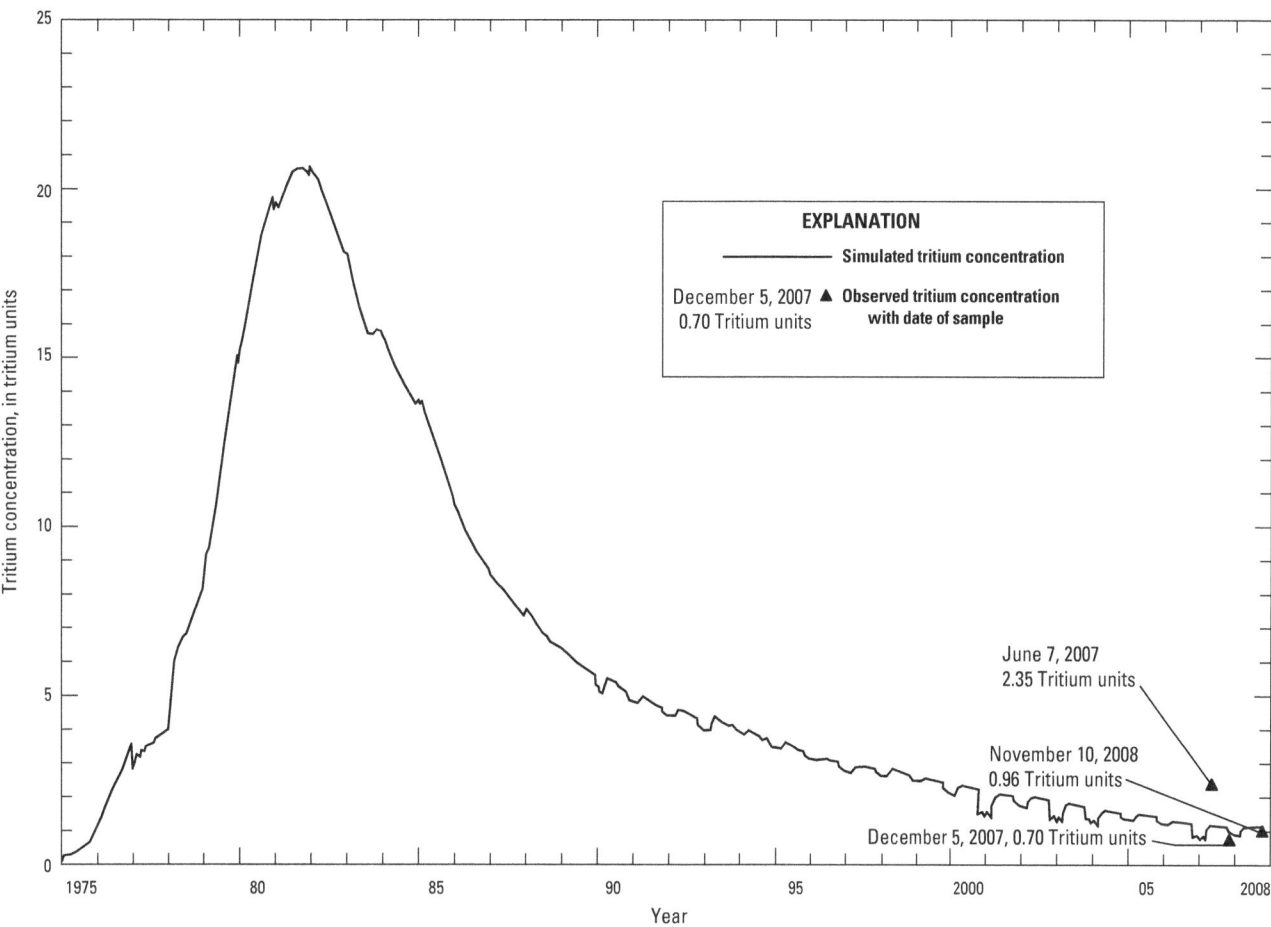

Figure 23. Tritium concentrations observed and simulated in the studied supply well in Albuquerque, New Mexico.

such period of low or no pumping from the SSW was simulated, and consequently no substantial flow was simulated out of the SSW into any model layer. The simulated variations in the young-water fractions among various model layers and the SSW borehole, such as those depicted and discussed for figure 29, therefore result from differing proportions of relatively young water migrating through the simulated aquifer material during different stress periods.

The borehole-flow logs (fig. 19), which were measured following a period of no pumping and water-level recovery, depict flow out of the SSW into the aquifer in discrete aquifer intervals. Although the details of these flows could not be simulated with the seasonal time discretization employed for this simulation, more detailed temporal discretization would enable more accurate simulation of periods of low pumping and no pumping, which could improve the simulation of mixing of borehole water returned to the aquifer with that in the surrounding aquifer, thereby enabling further analysis of the effects of borehole-return flow on the simulated age distribution of water surrounding the borehole. Quantification of the relative effects of borehole-return flow versus

aquifer-advective flow on the age-distribution variations could also help assess their relative importance as contaminant-transport pathways.

Simulated Fraction of Old Groundwater in the SSW

The seasonal variation in the fraction of deeper, "old" groundwater was simulated by a similar procedure to that previously described for the "young," post-1950 fraction. In the vicinity of the SSW, relatively high arsenic concentrations are associated with groundwater below a depth of 250 m (Laura Bexfield, USGS, written commun., 2012). To simulate the transport of this deeper fraction into the SSW, initial concentrations for the MT3DMS simulation were specified as zero within the upper model layers 1 through 37 that represent depths above about 250 m, and as 1 within the deeper model layers 38 through 45.

The simulated fraction of water originating below 250 m in the SSW (fig. 30) is a function of the simulated withdrawal

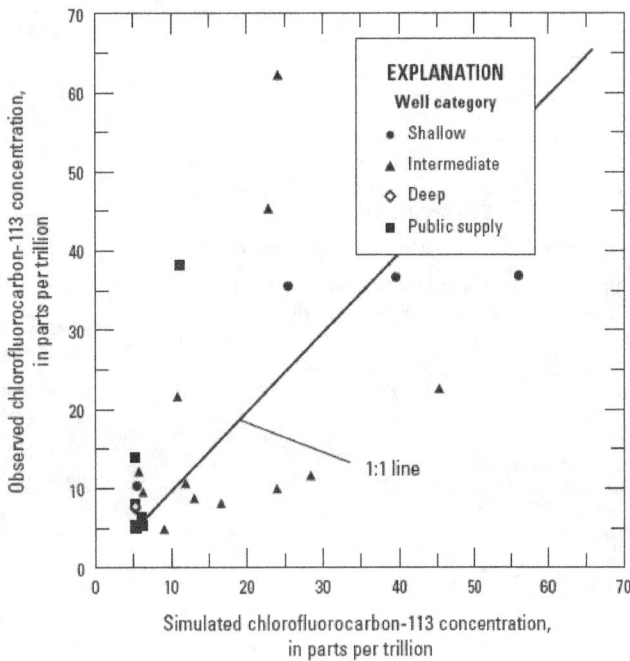

Figure 24. Relation between simulated and observed CFC-113 concentrations, Albuquerque, New Mexico.

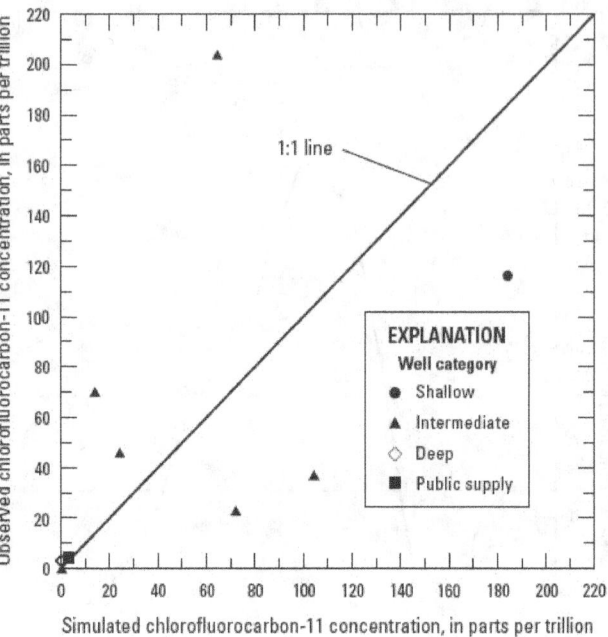

Figure 26. Relation between simulated and observed CFC-11 concentrations, Albuquerque, New Mexico.

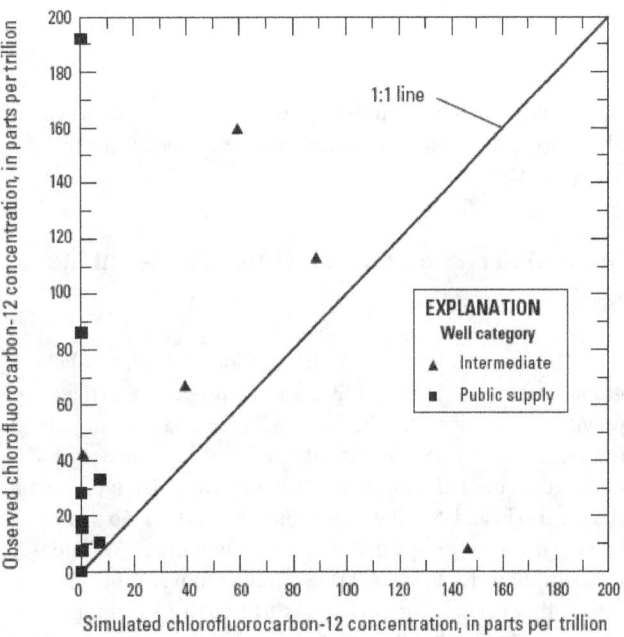

Figure 25. Relation between simulated and observed CFC-12 concentrations, Albuquerque, New Mexico.

rate from the SSW, which varies between the seasonal stress periods. The larger simulated variability in 2007 is caused by the finer temporal discretization used to simulate the period before and during an aquifer test and flow log in the SSW. During the summer seasons, when withdrawal rates are typically larger, a larger volume of the water pumped from the SSW is withdrawn from the upper, generally higher hydraulic-conductivity layers, resulting in smaller simulated fractions of "old" water during summer seasons. In contrast, smaller volumes withdrawn from the upper, generally higher hydraulic-conductivity layers during the winter result in a higher fraction of "old" water simulated in the SSW during the winter seasons. These simulated variations are consistent with the seasonal variability of observed arsenic concentrations in water samples from the SSW (Bexfield and others, 2012).

Simulated VOC Plume

The chlorinated solvents trichloroethylene (TCE) and cis-1,2-dichloroethylene (cis-1,2-DCE) and the gasoline oxygenate methyl tert-butyl ether (MTBE) were detected in low concentrations in groundwater produced by the SSW (Bexfield and others, 2012, table 8). Of the three known contaminant-source locations within the local study area (fig. 2), a Superfund site (U.S. Environmental Protection Agency, 2011) to the northwest of the SSW is upgradient along simulated flow paths and, therefore, is a possible source of contaminants to

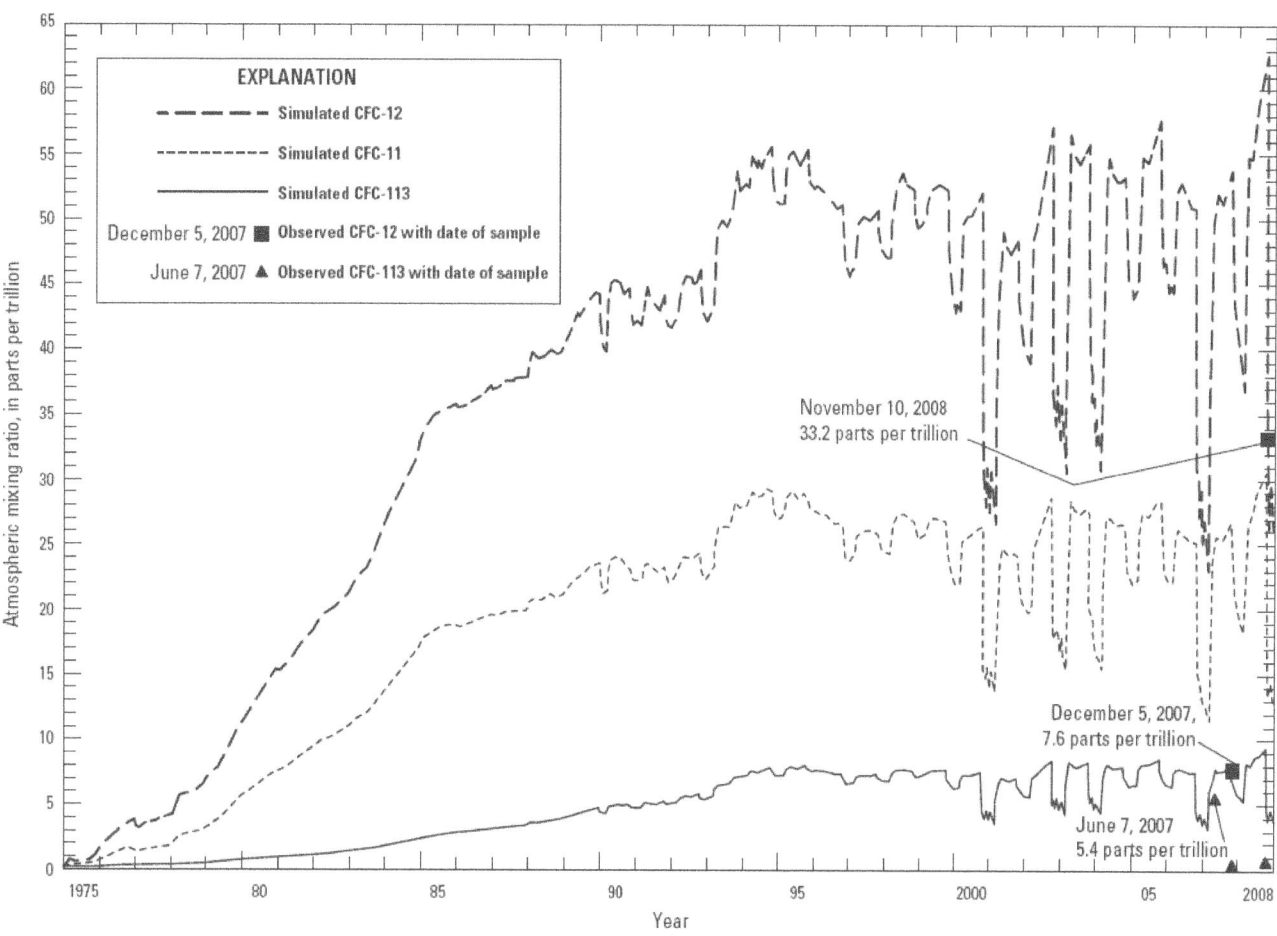

Figure 27. Simulated and observed CFC concentrations in the studied supply well in Albuquerque, New Mexico.

the SSW. The chlorinated solvents tetrachloroethene (PCE), TCE, and *cis*-1,2-DCE were detected in groundwater wells near this site and in soils on this site in 1989 (U.S. Environmental Protection Agency, 2001). The solvents likely leaked from onsite storage tanks associated with a laundry established in 1924 and (or) a dry-cleaning facility operated from 1940 through 1972. Chlorinated solvents are dense, non-aqueous phase liquids (DNAPLs) which generally sink through fresh groundwater until they dissolve into the groundwater and (or) adsorb onto the sedimentary aquifer material. Groundwater wells at and near the site may have provided preferential conduits for the chlorinated solvents to migrate to deeper aquifer intervals (Landin, 1999).

Flow paths from this possible contaminant-source site to the SSW were studied by using the particle tracking program MODPATH (Pollock, 1994), and the transient flow field was computed with the calibrated local-scale groundwater-flow model documented in this report. For these simulations, particles representing parcels of groundwater containing dissolved contaminants were distributed across the downgradient

faces of the top seven model cells that contain the relatively shallow groundwater well at the contaminant-source site. The particles were tracked forward in time along the simulated advective groundwater-flow paths until the simulated end date (Dec. 31, 2008), unless they either stopped at the perimeter boundary of the local-scale model or were stopped at a simulated well. Because the origination date and the duration of contaminant leakage were unknown, many simulation scenarios with different contaminant-release dates and simulated effective porosities were explored. Because the simulated rates of pumping from withdrawal wells and directions of groundwater flow changed during this transient simulation, the flow paths and fate of the particles were sensitive to the particle release date and the effective porosity specified in the particle-tracking algorithm, which controls the particle velocities. Particles released prior to about 1963 generally track the groundwater flow toward the south-southwest and sometimes toward withdrawal wells to the north-northeast. Subsequently, the drawdown of groundwater levels caused by withdrawals from public-supply wells inside and to the east of

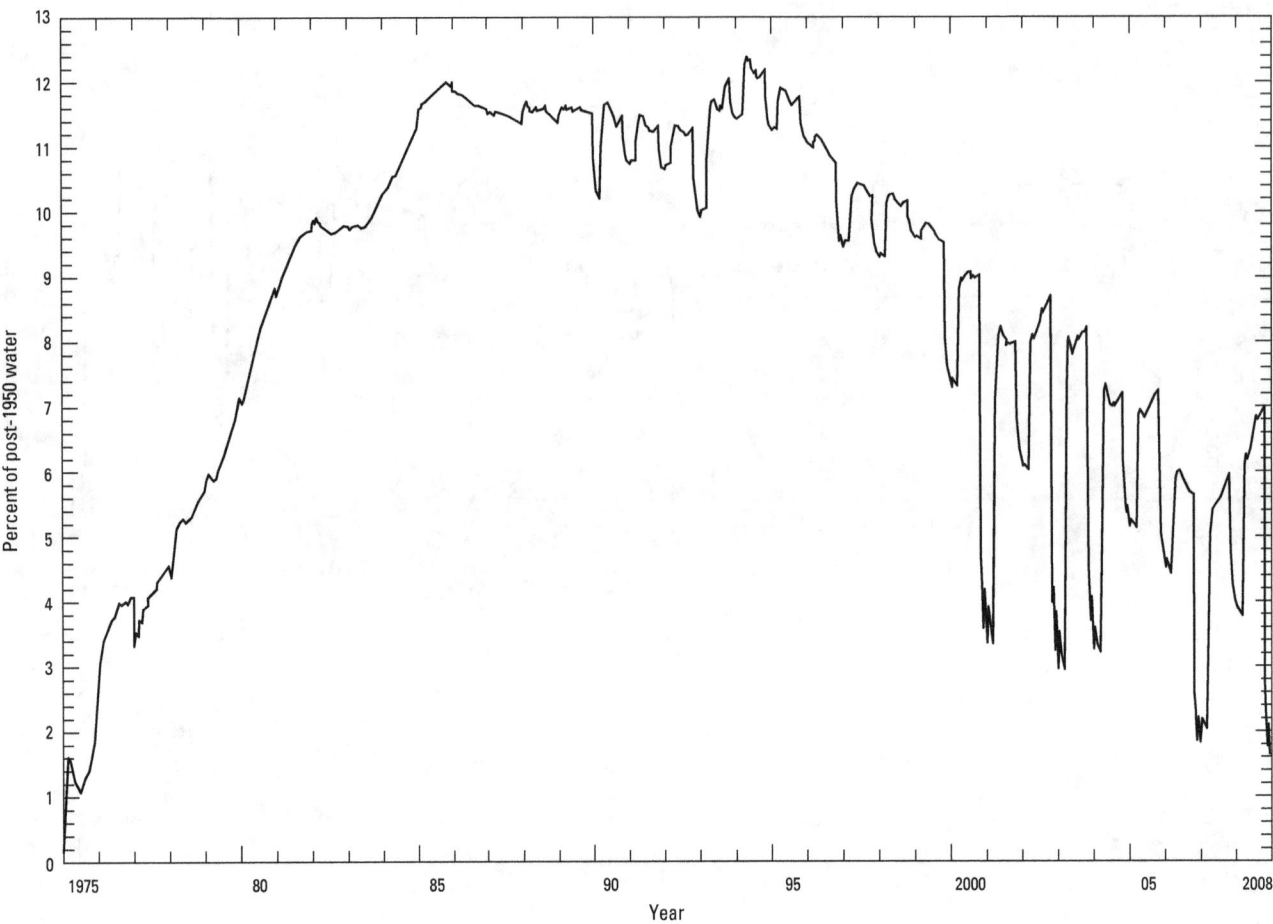

Figure 28.   Simulated percentages of post-1950 water in the SSW, Albuquerque, New Mexico, 1975–2005.

the local-scale model area direct the advective-transport flow paths toward the southeast. Because water-quality samples from locations south of the source were not obtained until after 1989, by which time the changed groundwater-flow direction might have flushed this area with clean water, the possibility of earlier release times cannot be ruled out. The preponderance of contaminated samples between the source and the SSW suggests that scenarios with relatively direct flow paths to the SSW reasonably represent the actual contaminant plume. Actual contaminant flow paths, however, are likely affected by "on and off" supply well pumping cycles that are not simulated with the seasonal time discretization of the flow simulation. For simplicity, two example simulations with different constant effective porosities will be discussed.

For the first example, the leakage was simulated by "releasing" particles from the source area once a year for 14 years (1975 until 1989) with the effective porosity (8 percent) that was estimated from calibration of the solute-transport model. A plot of the horizontal component of all particle flow paths in this scenario (fig. 31) illustrates that particles either exit the local-scale model domain in the easternmost

model column or are withdrawn through the well PSW1, and none arrive at the SSW. A groundwater "capture zone" created by withdrawals from PSW1, in particular, prevents the simulated flow paths of the contaminants from reaching the SSW in this example.

In the second example, particles that were released in 1976 were simulated with an effective porosity of 1.1 percent (fig. 32). In this case, some particles have sufficient velocity to avoid the capture zones of the intervening public-supply wells (PSW1 and PSW3) and are drawn into the SSW. Particle paths that terminate in the SSW have shallow flow paths near PSW3 and pass over the cells that contain the simulated screened interval of PSW3. Interestingly, the VOCs that were observed in the SSW and PSW1 were not observed in PSW3 (Bexfield and others, 2012, table 8). It is possible that the screened interval of PSW3 is isolated from shallower contaminant pathways. If that is the case, the pathways that terminate at the cell containing PSW3 in figures 31 and 32 may not be representative of the actual paths of VOCs from the depicted source.

Although the relatively small simulated effective porosity of 1.1 percent is likely less than the predominant effective

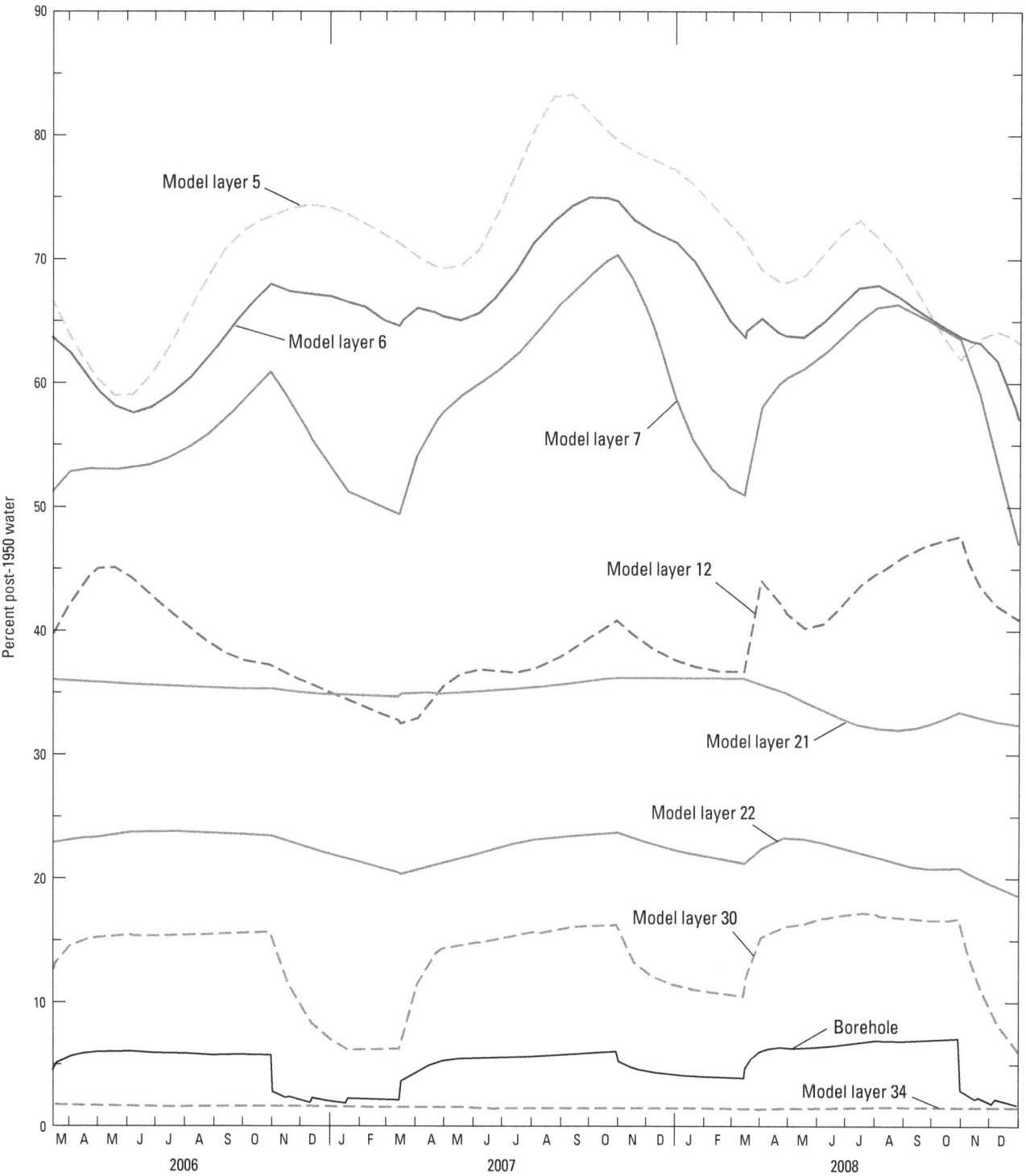

Figure 29.   Simulated percentages of post-1950 water in the aquifer surrounding the SSW, Albuquerque, New Mexico.

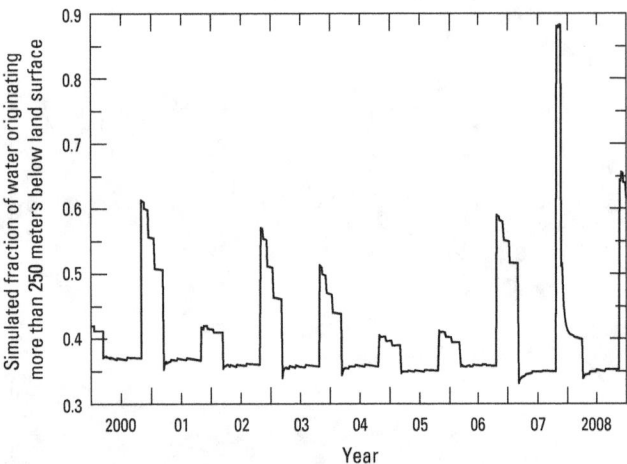

Figure 30. Simulated fraction of water originating more than 250 meters below land surface in the studied supply well in Albuquerque, New Mexico.

porosity in the aquifer, it may be representative for some preferential pathways, such as along sedimentary structures, that cannot be represented at a scale of spatial discretization in the model. In reality, an ensemble range of effective porosities likely exists in the aquifer.

Model Uncertainties and Limitations

In addition to the specification of boundary conditions, the simulated groundwater levels and transport of dissolved chemical tracers depend upon the specified spatial distribution of hydraulic properties controlling flow (hydraulic conductivity and storage) and transport (effective porosity). In this model, the spatial distribution of hydraulic conductivity has been simulated by one specific realization from a geostatistical hydrogeologic-facies model. Geostatistical properties defining characteristic spatial extents of the three hydrogeologic facies that compose the hydrogeologic-facies model, and the hydraulic properties associated with each of them, are uncertain. Although the specified distribution of the coarse, intermediate, and fine facies in model layers 3–34 does conform to the lithologies observed at the borehole-control locations, it is only intended to represent their relative abundance and the probable frequency of facies transitions at locations between those borehole locations. The amount of interconnection between the coarse facies, which are simulated by model cells with high hydraulic conductivities, in particular, likely has a substantial influence on the effectiveness of chemical-tracer and contaminant transport. Although the thickness and frequency of transition between facies were characterized in the vertical dimension at the borehole control points, the length, width, and interconnectedness of the coarse facies are not well constrained. Comparison of the longitudinal extents of the

interconnected coarse-facies model cells with the longitudinal extents of channel bottoms in the present-day Rio Grande and Rio Puerco suggests that the longitudinal coarse-facies interconnectedness is under-represented in the model. Future use of a depositional-process-based stratigraphic simulator could enable an improved representation of hydrostratigraphy in a similarly scaled model.

A regression calibration using a second hydrogeologic-facies realization resulted in slightly different parameter values. Because of the long transient-model runtimes, particularly for the multi-species transport model, it was not practical to calibrate the hydraulic-model parameters for each of the ten hydrogeologic facies (TPROGS) distributions. Forward groundwater-flow model runs and subsequent particle-tracking runs were made for all 10 realizations with an identical set of parameter values to ascertain if substantial differences in either simulated heads or transport pathways might be simulated with different realizations. Differences in simulated water levels between realizations were similar in magnitude to the residuals between the simulated and observed water levels for the calibrated realization.

Although the seasonal (230-day or 135-day) stress periods adequately simulate the different pumping rates which result in changing horizontal and vertical hydraulic gradients at the wellfield to a 6-km scale in the model, finer borehole-scale flows due to changing pumping rates over shorter periods are not well represented. The inter-day variability of withdrawals from the SSW and other public-supply wells, such as depicted in figure 7, likely causes differences in the hydraulic gradients and consequent flow into different vertical sections along the wellbore. The intra-borehole flow log (fig. 19) provided one measure of the vertical variability of flow into and out of the SSW following a transition from a non-pumping to pumping condition. Model testing using daily stress periods with different pumping rates has indicated that intra-borehole flows, such as those measured by the intra-borehole flow log, and the associated solute transport between model layers with different hydraulic heads can be simulated with the MNW2 package. Further simulations using daily stress period data could aid in discerning the relative importance of intra-borehole flow versus flow through the aquifer matrix in causing observed seasonal differences in chemical-tracer concentrations in the SSW.

The positive residuals for tritium and CFCs, with simulated values near zero and less than their observed counterparts, may indicate that some preferential transport pathways are not adequately simulated in the flow model. Although the TPROGs realizations may represent reasonable spatial variations in hydraulic properties, it is likely that some conductive vertical transport pathways exist in the aquifer at a finer spatial scale than that of the model discretization, and (or) should be represented with smaller effective porosity than the uniform value employed throughout the model domain. Additional well boreholes not represented in the model may also be short-circuiting flow though the aquifer matrix in some areas. Well boreholes that are simulated in the model may not adequately

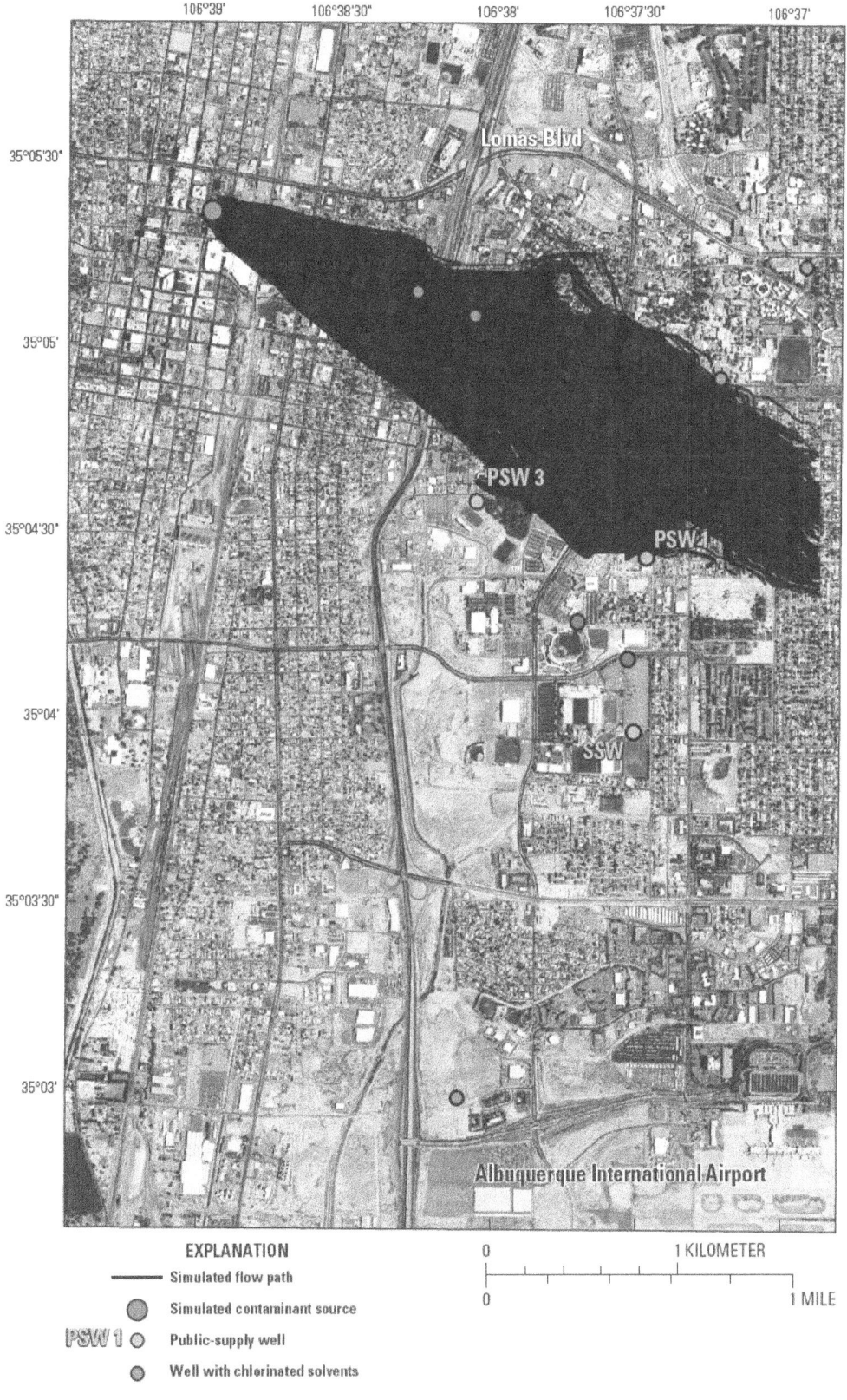

Figure 31. Flow paths of particles released from the contaminant source from 1975 until 1989 simulated with 8 percent effective porosity, public-supply wells, and monitoring wells with observed chlorinated solvents, Albuquerque, New Mexico.

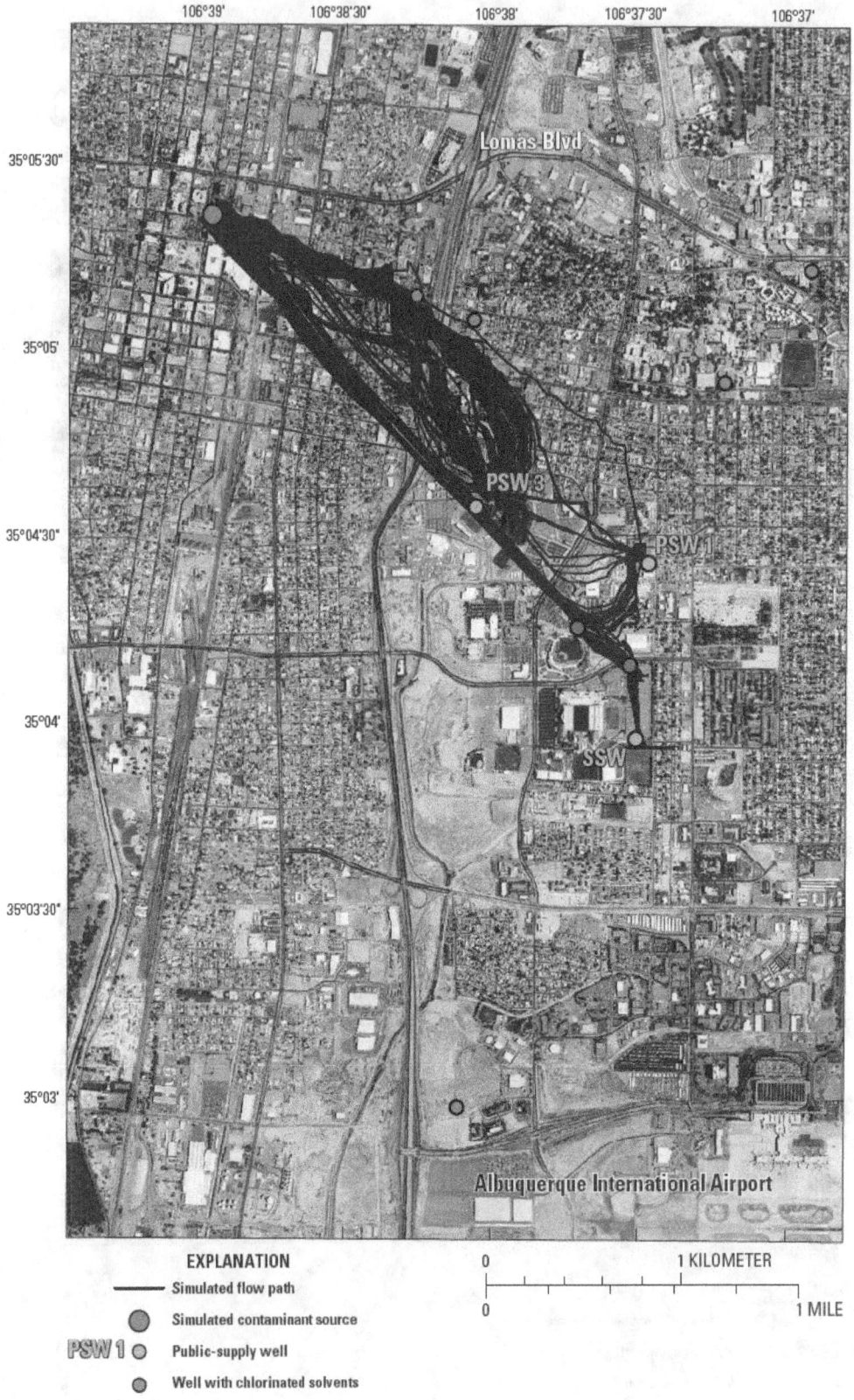

Figure 32. Flow paths of particles released from the contaminant source in 1976 simulated with 1.1 percent effective porosity, public-supply wells, and monitoring wells with observed chlorinated solvents, Albuquerque, New Mexico.

simulate the short-circuiting flow for a reason similar to that previously discussed for the SSW: the flow which may occur during short time intervals of non-pumping is not simulated by the seasonal stress periods, for which a higher pumping rate represents the average over the length of the stress period. A daily stress-period simulation is needed to further evaluate this possibility. Flow from the water table to the well screen may also occur through the borehole annular space around the casing of the SSW or other wells if they were imperfectly grouted or sealed, or if fractures in cement have developed over time. The possible effects of transport through this potential preferential flow path were not simulated.

Summary and Conclusions

This report documents the development and calibration of a model of transient groundwater flow and tracer-transport and selected particle-tracking simulations of a contaminant plume within the TANC local study area in Albuquerque, New Mexico. A 108-year transient groundwater-flow model of the 24-km^2 TANC study area was constructed and coupled to a modified version of the regional MODFLOW model of the MRGB with the local-grid refinement code MODFLOW-LGR2. Short-normal resistivity and lithologic logs from 14 wellbores and borehole cuttings from 4 wellbores were used to categorize the Tertiary alluvium that composes much of the zone of contribution to the public-supply wells into three hydrofacies. Ten hydrofacies-distribution realizations, each of which honored the borehole control data, were generated with the geostatistical software package TPROGS and were used to simulate a more complex hydraulic-conductivity distribution within 32 of the model layers that represented a 97-m-thick interval of the Tertiary Rio Puerco member of the Ceja Formation. Although all of these realizations were tested in various model runs, one realization was used for the final model calibration. Areas receiving recharge from sewer and water-distribution pipelines, infiltration from an unlined stormwater-diversion channel, and infiltration beneath ponds and irrigated vegetated areas at different times during the transient simulation were specified and parameterized so that the magnitude of recharge from each source could be estimated during model calibration.

The transport of five isotopic and chemical tracers (carbon-14, tritium, and three chlorofluorocarbon species) was simulated with the solute-transport code MT3DMS, which used the transient flow field computed by the local-scale MODFLOW simulation. The initial concentration condition for carbon-14 was interpolated from the regional transport simulation documented in the appendix. The concentrations specified for carbon-14 and tritium in recharge at the water table were delayed by 10 years with respect to their atmospheric concentrations in order to simulate 10 years of transport through the vadose zone. Measurements of water levels and tracer concentrations in the public-supply and other wells were used to calibrate parameters representing recharge, hydraulic conductivity, and effective porosity with the parameter-estimation code PEST. The uncertainty associated with each water level or concentration measurement was quantified and used to assign observation weights in the model-calibration regression. Values of some hydraulic-conductivity parameters were also constrained by measurements of borehole flow within the SSW.

The fit of simulated water levels to their measured values is graphically depicted with time series for wells with six or more observations and with a scatter plot of the relation between all water-level observations used in the model calibration and their simulated counterparts. Graphs of the relations between simulated and observed concentrations by observation-well-depth category for each of the five simulated species enable identification of possible model bias. The local-scale transport model simulates the set of carbon-14 observations within the local study area with greater accuracy and smaller bias than the regional-scale transport simulation documented in the appendix. The simulated positive residuals of tritium and CFCs, with simulated values near zero, may indicate that preferential-transport pathways are not adequately simulated in the flow model.

The simultaneous simulation of the concentrations of carbon-14 and the young-groundwater tracers (tritium and chlorofluorocarbons), which have different initial conditions, temporally changing input histories, and decay characteristics, constrained the simulation of the mixing of relatively recently recharged (post-1950) water with an older mixture containing groundwater that may have resided in the flow system for up to tens of thousands of years. The transient groundwater flow model simulates the effects of seasonal changes in pumping rates on the hydraulic gradients, and consequent transport rates, between different aquifer intervals and the screened intervals of the simulated withdrawal wells. Time-series plots illustrate the seasonal variability of simulated water levels and concentrations in the SSW, which are a consequence of increased groundwater withdrawals that typically occur during the summer stress periods. The relative fraction of groundwater originating in aquifer intervals above and below the studied supply well screened interval changes as a function of pumping rates and consequent hydraulic gradients and flow rates. Because these different aquifer intervals are more likely associated with either anthropogenic or natural contaminants in the local study area, the model is capable of simulating the change in susceptibility to contamination from either source as a function of withdrawal rates. The larger tracer concentrations simulated in the SSW during the summer seasons result from greater flow from the shallow aquifer intervals that contain relatively younger water, which generally contains higher concentrations of the tracers. The simulated percentage of post-1950 water is similarly greater during periods of greater groundwater withdrawal when larger hydraulic gradients exist between the water table and the SSW screened interval. These simulated results agree well with the observed seasonal tracer-concentration variability of water samples from the SSW and the independent mixing-model analysis in a companion report.

Seasonal variability in the fraction of water originating from deeper aquifer intervals near and below the bottom of the SSW screened interval, which is more likely associated with elevated arsenic concentrations, was also simulated for the SSW. The simulated fraction of this deeper-sourced, presumably older water is greater during periods of lower groundwater withdrawals, such as typically exist during the winter season. This pattern of simulated seasonal variability is in concordance with observations of higher arsenic concentrations in water samples from the SSW during the winter low-pumping season.

The observed temporal variability of measured concentrations may also be due, in part, to mixing of water in the aquifer resulting from the direct movement of water through the borehole (from intervals of higher head to lower head) during times when the well is not pumped. Because this borehole-flow mechanism was not adequately simulated with the seasonal stress periods, it remains uncertain whether flow through the aquifer or flow through the borehole is the dominant cause of vertical mixing in the vicinity of the SSW. Either mechanism likely results in the existence of zones of younger water, which are more likely to contain anthropogenic contaminants at greater depths than might exist without deep public-supply wells.

Particle tracking with MODPATH was employed to simulate the advective transport of dissolved contaminants from a known source located northwest of the SSW. Simulations using effective porosities of several percent, including that estimated by calibration of the solute-transport simulation (8 percent), and a wide range of contaminant-release times were characterized by capture of the simulated contaminated water by other public-supply wells located between the source and the SSW. A simulation with a smaller effective porosity of 1.1 percent, which may represent a particularly preferential pathway, simulated purely advective transport directly from the source into the SSW. Alternative scenarios with an early contaminant release time (around 1963) when groundwater flow was directed more southerly also simulate transport of contaminants into the SSW. Because water-quality samples from locations south of the source were not obtained until after 1989, by which time the changed groundwater-flow direction might have flushed the sampled area with clean water, this alternative scenario cannot be ruled out. Nevertheless, the preponderance of contaminated samples between the source and the SSW favors the scenario with a more direct path (and smaller simulated effective porosity) to the SSW.

Acknowledgments

The author acknowledges the assistance of people affiliated with the Albuquerque Bernalillo County Water Utility Authority (ABCWUA), the University of New Mexico (UNM), the Central New Mexico Community College (CNM), and the New Mexico Environment Department (NMED). Sean Connell of the New Mexico Bureau of Geology and Mineral Resources shared his extensive knowledge of the hydrogeology of the area and provided input on data collected as part of this study. Modeling of aquifer heterogeneity with the TPROGS software was performed by Nick Engdahl while attending UNM and working for the USGS. Nicole Thomas of the USGS described borehole cuttings, and Gordon Rattray and Eric Scherff analyzed borehole electric logs. Amjad Umari and Jesse Dickinson provided technical reviews, and Laura Bexfield and other members of the national TANC team provided advice.

References Cited

Alger, R.P., 1966, Interpretation of electric logs in fresh water wells in unconsolidated formations: Society of Professional Well Log Analysts Trans., Seventh Annual Logging Symposium, sec. CC, p. 1–25.

Bartolino, J.R., and Cole, J.C., 2002, Groundwater resources of the Middle Rio Grande Basin, New Mexico: U.S. Geological Survey Circular 1222, 132 p.

Bexfield, L.M., 2010, Conceptual understanding and groundwater quality of the basin-fill aquifer in the Middle Rio Grande Basin, New Mexico, in Thiros, S.A., Bexfield, L.M., Anning, D.W., Green, J.M., and McKinney, T.S., Conceptual understanding and groundwater quality of selected basin-fill aquifers in the Southwestern United States: U.S. Geological Survey Professional Paper 1781, p. 189–218.

Bexfield, L.M., Heywood, C.E., Kauffman, L.J., Rattray, G.W., and Vogler, E.T., 2011, Hydrogeologic setting and groundwater flow simulation of the Middle Rio Grande Basin regional study area, New Mexico, in section 2 of Eberts, S.M., ed., 2011, Hydrologic settings and groundwater flow simulations for regional studies of the transport of anthropogenic and natural contaminants to public-supply wells—Studies begun in 2004: U.S. Geological Survey Professional Paper 1737-B, p. 2-1–2-61.

Bexfield, L.M., Jurgens, B.C., Crilley, D.M., and Christenson, S.C., 2012, Hydrogeology, water chemistry, and transport processes in the zone of contribution of a public-supply well in Albuquerque, New Mexico, 2007–9: U.S. Geological Survey Scientific Investigations Report 2011–5182, 114 p.

Bjorklund, L.J., and Maxwell, B.W., 1961, Availability of ground water in the Albuquerque area, Bernalillo and Sandoval Counties, New Mexico: New Mexico State Engineer Technical Report 21, 117 p.

Carle, S.F., 1999, T-PROGS—Transition probability geostatistical software, version 2.1: Davis, Calif., University of California, 78 p.

Connell, S.D., 2006, Preliminary geologic map of the Albuquerque-Rio Rancho metropolitan area and vicinity, Bernalillo and Sandoval Counties, New Mexico: New Mexico Bureau of Geology and Mineral Resources Open-File Report 496, 2 pls.

Connell, S.D., Allen, B.D., and Hawley, J.W., 1998, Subsurface stratigraphy of the Santa Fe Group from borehole geophysical logs, Albuquerque area, New Mexico: New Mexico Geology, v. 20, no. 1, p. 2–7.

Croft, M.G., 1971, A method of calculating permeability from electric logs: U.S. Geological Survey Professional Paper 750-B, p. B265–B269.

Dickinson, J.E., James, S.C., Mehl, S.W., Hill, M.C., Leake, S.A., Zyvoloski, G.A., Faunt, C.C., and Eddebbarh, A.A., 2007, A new ghost-node method for linking different models and initial investigations of heterogeneity and nonmatching grids: Advances in Water Resources, v. 30, p. 1722–1736, doi:10.1016/j.advwatres.2007.01.004.

Doherty, John, 2004, PEST model-independent parameter estimation (5th ed.): Corinda, Australia, Watermark Numerical Computing, 336 p.

Feller, M.R., and Hester, D.J., 2001, SLEUTH Urban Landscape Change 2050 Calibration and Prediction datasets covering (16) 1:24,000-scale quadrangles within the Albuquerque, New Mexico Metropolitan Statistical Area, distributed to the New Mexico Bureau of Geology and Mineral Resources.

Grauch, V.J.S., and Connell, S.D., 2013, New perspectives on the geometry of the Albuquerque Basin, Rio Grande Rift, New Mexico—Insights from geophysical models of rift-fill thickness *in* Hudson, M.R., and Grauch, V.J.S., eds., New Perspectives on Rio Grande Rift Basins: From Tectonics to Groundwater: Geological Society of America Special Paper 494, p. 427–462, doi:10.1130/2013.2494(16).

Harbaugh, A.W., 2005, MODFLOW-2005, the U.S. Geological Survey modular ground-water model—The groundwater flow process: U.S. Geological Survey Techniques and Methods 6-A16 [variously paged].

Hazen, A., 1911, Discussion—Dams on sand foundations: Transactions of American Society of Civil Engineers, v. 73, p. 199.

Hinkle, S.R., Kauffman, L.J., Thomas, M.A., Brown, C.J., McCarthy, K.A., Eberts, S.M., Rosen, M.R., and Katz, B.G., 2009, Combining particle-tracking and geochemical data to assess public supply well vulnerability to arsenic and uranium: Journal of Hydrology, v. 376, p. 132–142.

Kernodle, J.M., McAda, D.P., and Thorn, C.R., 1995, Simulation of groundwater flow in the Albuquerque Basin, central New Mexico, 1901–1994, with projections to 2020: U.S. Geological Survey Water-Resources Investigations Report 94–4251, 114 p.

Konikow, L.F., Hornberger, G.Z., Halford, K.J., and Hanson, R.T., 2009, Revised multi-node well (MNW2) package for MODFLOW ground-water flow model: U.S. Geological Survey Techniques and Methods 6–A30, 67 p.

Landin, B.K., 1999, Background investigation report, Fruit Avenue plume, Albuquerque, New Mexico, CERCLIS # NMD986668911: New Mexico Environment Department, Groundwater Quality Bureau, Superfund Oversight Section, Volume I of III, prepared for the U.S. Environmental Protection Agency, Region 6 [variously paged].

McAda, D.P., and Barroll, P., 2002, Simulation of groundwater flow in the Middle Rio Grande Basin between Cochiti and San Acacia, New Mexico: U.S. Geological Survey Water-Resources Investigations Report 02–4200, 81 p.

Mehl, S.W., and Hill, M.C., 2005, MODFLOW-2005, the U.S. Geological Survey modular ground-water model—Documentation of shared node local grid refinement (LGR) and the boundary flow and head (BFH) package: U.S. Geological Survey Techniques and Methods 6–A12, 68 p.

Mehl, S.W., and Hill, M.C., 2013, MODFLOW–LGR—Documentation of Ghost Node Local Grid Refinement (LGR2) for Multiple Areas and the Boundary Head and Flow (BFH2) Package: U.S. Geological Survey Techniques and Methods 6-A44, (in press).

Michel, R.L., 1989, Tritium deposition over the continental United States, 1953–83, *in* Atmospheric Deposition: Oxfordshire, England, International Association of Hydrological Sciences, p. 109–115.

Neuman, S.P., 1974, Effect of partial penetration on flow in unconfined aquifers considering delayed gravity response: Water Resources Research, v. 10, no. 2, p. 303–312.

New Mexico Environmental Finance Center, 2006, Review of leak and repair data, phase 1 report prepared for the Albuquerque Bernalillo County Water Utility Authority, 91 p., accessed May 2009 at *http://nmefc.nmt.edu/ documents/WUA_Phase_1_Report_Final.pdf.*

Paschke, S.S., ed., 2007, Hydrogeologic settings and ground-water flow simulations for regional studies of the transport of anthropogenic and natural contaminants to public-supply wells—Studies begun in 2001: U.S. Geological Survey Professional Paper 1737-A, 244 p.

Peterson, M.A., 1992, A summary history of AMAFCA, 17 p., accessed May 2012 at *http://www.amafca.org/documents/summaryhistory.pdf.*

Plummer, L.N., Bexfield, L.M., Anderholm, S.K., Sanford, W.E., and Busenberg, E., 2004, Geochemical characterization of ground-water flow in the Santa Fe Group aquifer system, Middle Rio Grande Basin, New Mexico: U.S. Geological Survey Water-Resources Investigations Report 03–4131, 395 p.

Poeter, E.P., and Hill, M.C., 1998, Documentation of UCODE, a computer code for universal sensitivity inverse modeling: U.S. Geological Survey Water-Resources Investigations Report 98–4080, 116 p.

Poeter, E.P., Hill, M.C., and Banta, E.R., 2005, UCODE_2005 and six other computer codes for universal sensitivity analysis, calibration, and uncertainty evaluation: U.S. Geological Survey Techniques and Methods, book 6, chap. 11, 283 p.

Pollock, D.W., 1994, User's guide for MODPATH/MOD-PATH-PLOT, version 3—A particle tracking post-processing package for MODFLOW, the U.S. Geological Survey finite-difference ground-water flow model: U.S. Geological Survey Open-File Report 94–464, [variously paged].

Sanford, W.E., Plummer, L.N., McAda, D.P., Bexfield, L.M., and Anderholm, S.K., 2004, Use of environmental tracers to estimate parameters for a predevelopment ground-water-flow model of the Middle Rio Grande Basin, New Mexico: U.S. Geological Survey Water-Resources Investigations Report 03-4286, 102 p.

Shepherd, R.G., 1989, Correlations of permeability and grain size: Ground Water, v. 27, no. 5, p. 21–28.

Thorn, C.R., McAda, D.P., and Kernodle, J.M., 1993, Geo-hydrologic framework and hydrologic conditions in the Albuquerque Basin, central New Mexico: U.S. Geological Survey Water-Resources Investigations Report 93–4149, 106 p.

U.S. Environmental Protection Agency, 2001, Record of decision, Fruit Avenue Plume site, Albuquerque, New Mexico, CERCLIS # NMD986668911: U.S. Environmental Protection Agency, Region 6, Superfund Division [variously paged].

U.S. Environmental Protection Agency, 2011, Fruit Avenue Plume (Bernalillo County), New Mexico, EPA ID# NMD986668911, accessed November 2011 at *http://www.epa.gov/earth1r6/6sf/pdffiles/0604068.pdf.*

Zheng, C., and Wang, P.P., 1999, MT3DMS—A modular three-dimensional multispecies transport model for simulation of advection, dispersion, and chemical reactions of contaminants in groundwater systems—Documentation and user's guide: Prepared for U.S. Army Corps of Engineers, monitored by U.S. Army Engineer Research and Development Center (Contract report ; SERDP-99-1), 221 p.

Zheng, C., 2010, MT3DMS v. 5.3, Supplemental user's guide: Tuscaloosa, Alabama, Department of Geological Sciences, University of Alabama, 51 p.

Appendix

Appendix   Simulation of the regional distribution of carbon-14 concentration.

The radiocarbon age of dissolved inorganic carbon in groundwater can be used to estimate the aquifer-residence time, or the time since recharge to the aquifer system, of a groundwater sample. The carbon-14 concentrations in numerous groundwater samples from the Middle Rio Grande Basin (MRGB) were measured for the NAWQA TANC study (Bexfield and others, 2011) as well as previous studies (Plummer and others, 2004). Observations of carbon-14 concentration in the local-scale model area (table 4) were used to estimate a value of regional effective porosity in a regional-scale carbon-14 transport simulation and to calibrate hydraulic conductivities and effective porosity in the local-scale groundwater-flow and solute-transport simulations. The objective of the regional carbon-14 transport simulation was to provide a reasonable representation of the predevelopment (steady-state) carbon-14 distribution in the area of the local model, which, in turn, was used to specify the initial carbon-14 condition for the local-scale transient carbon-14 transport simulation described in this report.

The transport code MT3DMS numerically solves the transient advection-dispersion equation:

$$\frac{\partial\left(\theta C^k\right)}{\partial t} = \frac{\partial}{\partial x_i}\left(\theta D_{ij}\frac{\partial C^k}{\partial x_j}\right) - \frac{\partial}{\partial x_i}\left(\theta v_i C^k\right) + q_s C_s^k + R_n \quad , (1\text{-}1)$$

where

θ	is	the effective porosity,
C^k	is	the concentration of the dissolved chemical species k,
t	is	time,
x	is	distance,
i and j	are	indices referencing orthogonal directions in a Cartesian coordinate system,
D	is	the hydrodynamic dispersion tensor,
v_i	is	the seepage velocity,
q_s	is	the volumetric flow rate of fluid sources and sinks,
C_s^k	is	the concentration of dissolved chemical species k in the fluid sources or sink, and
R_n	is	a chemical reaction term.

The radioactive decay of dissolved carbon-14 (represented by the term R_n in equation 1-1) was simulated as a first-order irreversible reaction without sorption by using a decay constant representing the 5,730-year half-life in the Chemical Reaction Package of MT3D.

The regional transport model simulates the regional distribution of carbon-14 concentration, which primarily depends upon the initial carbon-14 concentration of water when it recharges the groundwater system and the elapsed time of transport between recharge and the sampling event. Because some numerical dispersion always occurs in a numerical solution of the solute transport equation (eq. 1-1) and adequate criteria for specification of aquifer system dispersivities were not available, additional dispersive mixing was not simulated.

Therefore, the regional transport model primarily simulates advective transport, with some numerical dispersion.

The spatial and temporal discretizations of the regional MT3D transport simulation are identical to the revised regional flow model, which enables use of the MODFLOW-2005 computed inter-cellular flows to define the advective term of the transport equation (eq. 1-1). Corresponding to the MODFLOW simulation, the MT3DMS transport simulation consists of steady-state and transient stress periods. The predevelopment carbon-14-concentration distribution was simulated in an initial steady-state stress period solved with the fully implicit finite-difference solution option of MT3DMS with upstream weighting.

Concentrations were specified at model boundaries that simulate recharge to the groundwater system along rivers (such as the Rio Grande and Rio Puerco), mountain-front areas, and, in the transient simulation only, beneath leaky water and sewer pipes in the Albuquerque area. The carbon-14 concentration of recharge water in the MRGB is likely near 100 percent modern carbon (pmc); accordingly, this value was assigned to model cells containing river or recharge boundaries with the Sink & Source Mixing Package of MT3DMS. However, because vadose-zone geochemical reactions can lower the initial carbon-14 concentration of recharge water, these concentrations were parameterized to enable their estimation during model calibration. Because groundwater underflow into the regional model area is thought to be old, a carbon-14 concentration of zero was specified for such boundaries.

Calibration of the Regional-Scale Carbon-14 Transport Simulation

The values of parameters representing a uniform effective porosity and initial carbon-14 concentration of water in rivers and other recharge sources were estimated with UCODE (Poeter and others, 2005) by minimizing the objective function:

$$\sum_{i=1}^{N} \omega_i \left(C_i - C'_i\right)^2 \quad , \qquad (1\text{-}2)$$

where

C_i	is	the measured carbon-14 concentration for observation i,
C'_i	is	the simulated-equivalent carbon-14 concentration observation for i,
ω_i	is	the weight applied to the squared difference of observation i and its simulated equivalent, and
N	was	35, which is the number of carbon-14 observations in the local-scale study area that were used in the regression.

The weight (ω_i) assigned to each carbon-14 observation was equal to the inverse of the calculated total error variance for that observation. The total observation error variance was the sum of the laboratory, or "measurement," error variance and the error variance introduced in the geochemical modeling by Bexfield and others (2012),which introduced the majority of the uncertainty. The estimated variance for each modeled carbon-14 observation was based on two assumptions: (1) a normal distribution and (2) 90-percent confidence that the true value lies between the measured value and the "best" geochemically modeled value. Stated differently, the magnitude of the uncertainty interval introduced from the geochemical modeling was estimated with 90-percent confidence to be the difference of the corrected and measured values.

In each iteration of the UCODE regression, the parameter values for a combined steady-state and transient regional MT3D transport simulation were updated. A uniform effective porosity of 18 percent and a recharge-source carbon-14 concentration of 100 pmc produced the optimal fit between observed carbon-14 concentrations and their corresponding values simulated with the transient transport model.

Simulated Regional Carbon-14 Concentrations

The steady-state regional distribution of carbon-14 concentrations simulated within the top layer of the transport model, which represents water at depths from zero to 14 meters below the simulated steady-state water table, is depicted in figure 1-1. Areas of high simulated carbon-14 concentration result from the shorter times of groundwater transport since recharge at the base of the mountains along the eastern basin margin, the Rio Puerco, and losing reaches of the Rio Grande. The simulated distribution is comparable to that mapped by Plummer and others (2004, fig. 77) from measured carbon-14 concentrations. Simulated carbon-14 concentrations in deeper model layers are smaller, representing generally increasing groundwater age with depth, and have similar spatial variability. The area encompassed by the local-scale model domain described in the main report is also depicted in figure 1-1. The steady-state carbon-14 concentrations simulated with the regional transport model described in this appendix were used to specify the initial carbon-14 concentrations for the local-scale model described in the main report. The non-random fit of carbon-14 concentrations simulated with the regional-scale transport model to concentrations observed in the local study area (fig. 1-2) suggests that the regional model over-predicts the age (under-predicts the carbon-14 concentration) of relatively young groundwater samples and under-predicts the age (over-predicts the carbon-14 concentration) of older groundwater samples in this area.

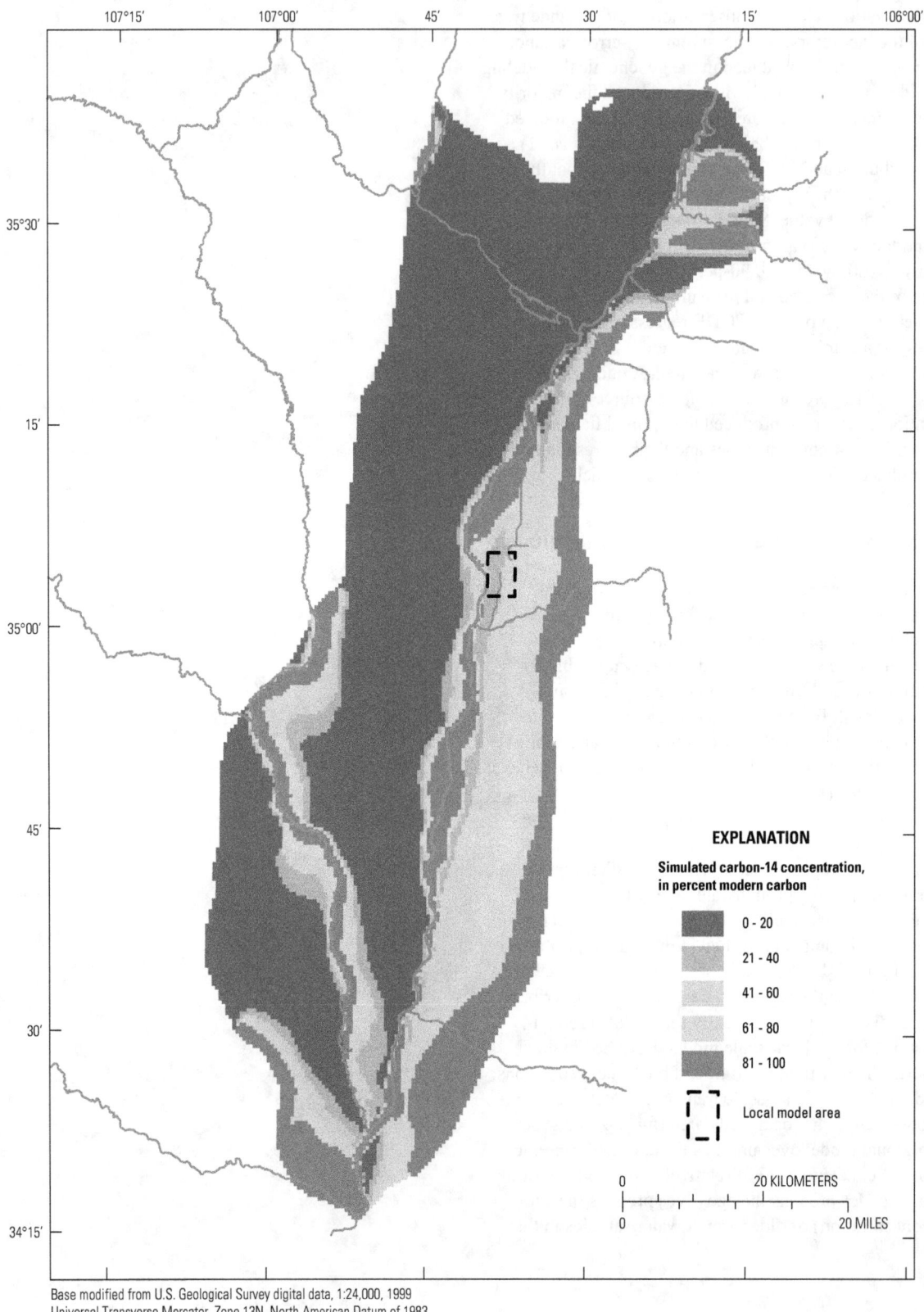

Base modified from U.S. Geological Survey digital data, 1:24,000, 1999
Universal Transverse Mercator, Zone 13N, North American Datum of 1983.

Figure 1-1.   Carbon-14 concentrations simulated in the top layer of the Albuquerque Basin regional transport model.

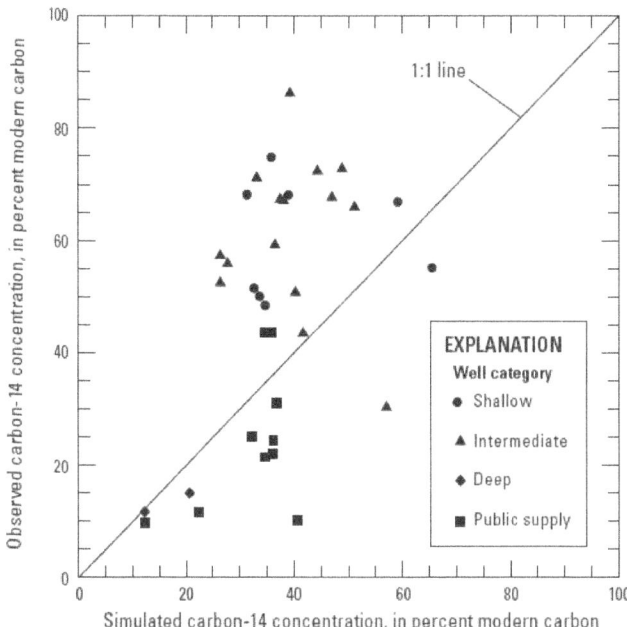

Figure 1-2.   Relation between observed carbon-14 concentrations and concentrations simulated with the Albuquerque Basin regional transport model.

www.ingramcontent.com/pod-product-compliance
Lightning Source LLC
Chambersburg PA
CBHW081615170526
45166CB00009B/2968